Military Necessity and Just War Statecraft

This book analyzes the concept of military necessity and just war thinking and argues that it should be seen as a vital moral principle for leaders.

The principle of *military necessity* is well understood in the manuals of modern militaries and is recognized in the war convention. It is the idea that battlefield commanders should make every effort to win on a local battlefield, within legal means, and using proportionate and discriminating weapons and tactics. Every legal textbook on war includes military necessity as a foundational principle within the *jus in bello* (ethics of fighting war) alongside principles of proportionality and distinction, and it is taught in every Western military academy. Even the International Committee of the Red Cross lauds the concept as a cardinal principle of warfare. However, unlike legal scholarship, one can pick up a book by almost any just war thinker in philosophy, theology, or the social sciences, and the concept is missing altogether in their literature. This volume returns military necessity to just war thinking and lays out the argument for doing so. Each contributor taps into one of the many dimensions of military necessity, such as its relationship to *jus ad bellum* (ethics of going to war) categories (e.g., right intention), its relationship to *jus in bello* categories, or its application in foreign policy and military doctrine. Case studies in the book point out the practical moral dimensions of military necessity in cases from the targeted killing of terrorists to battlefield decisions that led to the use of the atomic bomb at Hiroshima.

This book will be of interest to students of just war theory, military ethics, statecraft, and international relations.

Eric Patterson is scholar-at-large and former dean of the School of Government at Regent University. He is author or editor of 20 books, including, most recently, *Just War and Christianity: A Concise Introduction* (2023) and *Just American Wars* (2019).

Marc LiVecche is the McDonald Distinguished Scholar of Ethics, War, and Public Life at *Providence: A Journal of Christianity & American Foreign Policy* and serves as a non-resident research fellow at the College of Leadership and Ethics in the U.S. Naval War College. He is author of *The Good Kill: Just War and Moral Injury* (2021).

War, Conflict and Ethics
Series Editors: Michael L. Gross, University of Haifa
and James Pattison, University of Manchester

Ethical judgments are relevant to all phases of protracted violent conflict and inter-state war. Before, during, and after the tumult, martial forces are guided, in part, by their sense of morality for assessing whether an action is (morally) right or wrong, an event has good and/or bad consequences, and an individual (or group) is inherently virtuous or evil. This new book series focuses on the morality of decisions by military and political leaders to engage in violence and the normative underpinnings of military strategy and tactics in the prosecution of the war.

Distributing the Harm of Just Wars
In Defence of an Egalitarian Baseline
Sara Van Goozen

Moral Injury and Soldiers in Conflict
Political Practices and Public Perceptions
Tine Molendijk

The Empathetic Soldier
Kevin R. Cutright

Law, Ethics and Emerging Military Technologies
Confronting Disruptive Innovation
George Lucas

Ethics at War
How Should Military Personnel Make Ethical Decisions?
Deane-Peter Baker, Rufus Black, Roger Herbert and Iain King

Military Necessity and Just War Statecraft
The Principle of National Security Stewardship
Edited by Eric Patterson and Marc LiVecche

For more information about this series, please visit: https://www.routledge.com/
War-Conflict-and-Ethics/book-series/WCE

Military Necessity and Just War Statecraft

War Statecraft

The Principle of National Security Stewardship

Edited by
Eric Patterson and Marc LiVecche

Routledge
Taylor & Francis Group

LONDON AND NEW YORK

First published 2024
by Routledge
4 Park Square, Milton Park, Abingdon, Oxon OX14 4RN

and by Routledge
605 Third Avenue, New York, NY 10158

Routledge is an imprint of the Taylor & Francis Group, an informa business

British Library Cataloguing-in-Publication Data
A catalogue record for this book is available from the British Library

Library of Congress Cataloging-in-Publication Data
Names: Patterson, Eric, 1971– editor. | LiVecche, Marc, editor.
Title: Military necessity and just war statecraft : the principle of national security stewardship / edited by Eric Patterson and Marc LiVecche.
Description: Abingdon, Oxon ; New York, NY : Routledge, 2024. |
Series: War, conflict and ethics | Includes bibliographical references and index.
Identifiers: LCCN 2023038435 (print) | LCCN 2023038436 (ebook) |
ISBN 9781032487090 (hardback) | ISBN 9781032487120 (paperback) |
ISBN 9781003390398 (ebook)
Subjects: LCSH: Just war doctrine. | National security—Moral and ethical aspects. | Necessity (International law)
Classification: LCC U21.2 .M544 2024 (print) | LCC U21.2 (ebook)
LC record available at https://lccn.loc.gov/2023038435
LC ebook record available at https://lccn.loc.gov/2023038436

ISBN: 978-1-032-48709-0 (hbk)
ISBN: 978-1-032-48712-0 (pbk)
ISBN: 978-1-003-39039-8 (ebk)

DOI: 10.4324/9781003390398

Typeset in Times New Roman
by codeMantra

Contents

Notes on Contributors

Christian Nikolaus Braun is a lecturer at the Defence Studies Department, King's College, London. Previously, he was a Radboud Excellence Initiative Fellow at Radboud University and a senior lecturer at the Royal Military Academy, Sandhurst. Christian's primary area of research is the ethics of war and peace. It is his intellectual ambition to bring to bear the wisdom of the just war tradition on the changing character of war. Christian's work has been published in leading academic journals, including *Ethics & International Affairs*, *Global Studies Quarterly*, *International Relations*, and *International Theory*. He is also the author of *Limited Force and the Fight for the Just War Tradition* (2023).

Louis Bujnoch was awarded his PhD in 2020 from the University of Glasgow for his dissertation examining the concept of necessity in war and conflict. It traces different understandings of necessity relating to war and conflict across multiple discourses and knowledge communities, such as international law, military strategy, and Just War thought. His main research interest lies at the intersection of the ethics of war, normative (international) politics, and the use of force and coercive means by states. Beyond this, he is more broadly interested in international political thought and the role of technology in (international) politics. He currently works as a teaching fellow at the Department of Politics and International Relations, University of Glasgow, Scotland.

Pedro Erik Carneiro holds a PhD in International Relations and a master's degree in Economics both from the University of Brasília (Brazil). He is author of two books related to Catholic Doctrine in Brazil; the first is on just war theory, *Teoria e Tradição da Guerra Justa: do Império Romano ao Estado Islâmico* (2016), and the second is on Catholic ethics for economics, *Ética Católica para Economia: Bíblia, Teólogos e Ciência Econômica* (2019). Regarding war issues, he is the author of articles related to Islamic terrorism, including "God and the Dystopias" (2018) and "Trying to Catch the Deluge: Shar'iah, Terrorism, and Religious Freedom" (2014), and a chapter "Tear Down This Wall: The Tri-Border Area Terrorism" (2012). He also has articles on finance and the environment. He is a professor of Economics at Centro Universitário Unieuro (Brasília, Brazil) and has been working at the Brazilian Ministry of Finance since 2001.

J. Daryl Charles is an affiliate scholar of the John Jay Institute and has served as the Acton Institute affiliated scholar in Theology and Ethics. He is a contributing editor at *Providence: A Journal of Christianity & American Foreign Policy.* He is author, coauthor, or editor of 21 books, including *Just War and Christian Traditions* (2022, with Eric Patterson), *America and the Just War Tradition: A History of U.S. Conflicts* (2019, with Mark David Hall), *The Just War Tradition: An Introduction* (2012, with David D. Corey), *War, Peace, and Christianity* (2010, with Timothy J. Demy), and *Between Pacifism and Jihad* (2005).

Jonathan French Flint holds a BA in International Studies from the Open University in the United Kingdom. His postgraduate education was at the University of Hull, where he earned an MA in Strategy and International Security and a PhD in Politics, focusing on strategic theory. His advisors at Hull included two former students of Professor Colin Grey. He has taught undergraduate and graduate courses in the UK and at Case Western Reserve University in Cleveland, Ohio. He has presented at conferences on both sides of the Atlantic, including the 2016 Euro-ISME (International Society for Military Ethics) conference in Oslo, Norway, US-ISME, and the US Army's Command and General Staff College's annual Ethics Symposium at Ft. Leavenworth, Kansas. He has published on military ethics with the ICRC. He sometimes appears in media as a subject matter expert on the war in Ukraine, including on war crimes allegations, as well as military and strategic affairs.

Shannon E. French joined the faculty at Case Western Reserve University (CWRU) in 2008 after teaching for 11 years at the US Naval Academy, Annapolis, where she was associate chair of the division of Leadership, Ethics, and Law. At CWRU, she is the Inamori professor of Ethics, professor of Philosophy and Law, director of the Inamori International Center for Ethics and Excellence, and a director of the first MA program in Military Ethics in the United States. Her primary research fields are military ethics and ethics and emerging technology. Her publications include *The Code of the Warrior* (third edition forthcoming), several edited volumes, and numerous chapters and articles, including "Distinction and Civil Immunity," "Artificial Intelligence in Military Decision-Making," "War and Technology: Should Data Decide Who Lives, Who Dies?" "Neuroethics, Dehumanization, and the Prevention of Moral Injury," and "Military Neuro-Interventions: Solving the Right Problems for Ethical Outcomes." She is an ELSI (Ethical, Legal, and Social Issues) consultant for the Institute for Defense Analysis (IDA) and the RAND Corporation and a member of the ethics board for the ACM (Association for Computing Machinery). Since 2017, she has served as the General Hugh Shelton Distinguished Chair in Ethics for the CGSC (Command and General Staff College) Foundation. She is editor-in-chief of the *International Journal of Ethical Leadership,* an associate editor for the *Journal of Military Ethics,* on several other editorial boards and is active in the European Chapter of the International Society for Military Ethics (Euro-ISME). She is also a senior research fellow for the Simons Center for Ethical Leadership and Interagency Cooperation.

Amos N. Guiora is professor of Law at the S.J. Quinney College of Law, the University of Utah. He is a distinguished fellow at the Consortium for the Research and Study of Holocaust and the Law at Chicago-Kent College of Law, and a distinguished fellow and counselor at the International Center for Conflict Resolution, Katz School of Business, University of Pittsburgh. Professor Guiora is on the board of the Lauren McClusky Foundation. He is the recipient of the University of Utah's Distinguished Faculty Service Award. For the past ten years, Guiora has been researching-writing-lecturing on the question of bystanders (originally in the Holocaust) resulting in his books, *The Crime of Complicity: The Bystander in the Holocaust* (2017) and *Armies of Enablers: Survivor Stories of Complicity and Betrayal in Sexual Assaults* (2020). His most recent article, "Holding Enablers of Child Sexual Abuse Accountable: The Case of Jeremy Bell" (2023), addresses the role of enablers in sexual assault of children. He directs the SJ Quinney College of Law Bystander Initiative. He has an AB in History from Kenyon College, a JD from Case Western Reserve University School of Law, and a PhD from Leiden University.

Joshua Hastey received his PhD from Old Dominion University (ODU). His dissertation examines the relationship between power shifts and China's willingness to attempt revisions to territorial status quo. He teaches and writes on international relations and security studies, with special interests in Indo-Pacific security, gray zone/hybrid strategies, and just war thought. He is the author of *China, Faits Accomplis, and the Contest for Asia* (2023). Hastey has published book chapters and articles with Bristol University Press, *Journal of Strategic Security*, *Providence: A Journal of Christianity & American Foreign Policy*, and *World Policy Journal* and served as a guest speaker/moderator for classes, panels, and podcasts related to international politics and U.S. grand strategy. He also serves as an adjunct professor of strategy for the U.S. Naval War College's Fleet Seminar Program.

Pauline Shanks Kaurin is professor and Admiral James B. Stockdale Chair in Professional Military Ethics. She holds a PhD in Philosophy from Temple University and specializes in military ethics, "just war theory," and philosophy of law and applied ethics. She is author of *On Obedience: Contrasting Philosophies for Military, Community and Citizenry* and *Achilles Goes Asymmetrical: The Warrior, Military Ethics and Contemporary Warfare*. She served as a contributor for *The Strategy Bridge* and has published in *War on the Rocks*, *Clear Defense*, *The Wavell Room*, *Grounded Curiosity*, *Newsweek*, and *Just Security*. She is a member and serves on the board of directors for the Military Writers Guild.

Marc LiVecche is the McDonald Distinguished Scholar of Ethics, War, and Public Life at *Providence: A Journal of Christianity & American Foreign Policy* and a non-resident research fellow at the U.S. Naval War College. He is the author of *The Good Kill: Just War and Moral Injury* (2021) and the *Moral Horror: A Just War Defense of the Bombing of Hiroshima* (2024). He has published numerous articles and chapters across a variety of journals and edited volumes and has taught and lectured widely both internationally and domestically. He is active in a number of academic and professional associations, including the International Society of Military Ethics. He previously taught at the U.S. Naval Academy where he was a McCain Fellow at the Stockdale Center for Military Ethics and was McDonald Visiting Scholar at the McDonald Centre for Theology, Ethics, and Public Life, at Christ Church, Oxford.

David Luban is distinguished university professor and professor of Law and Philosophy at Georgetown University. Since 2013, he has also served as Class of 1965 Distinguished Chair in Ethics at the U.S. Naval Academy's Stockdale Center for Ethical Leadership. In addition to legal and military ethics, he writes on international criminal law, national security, and just war theory. Most recently he has authored *Torture, Power, and Law* (2014) and is a coauthor of *International and Transnational Criminal Law*, fourth edition (2018). He is a member of the American Academy of Arts and Sciences, and a distinguished fellow of the National Institute of Military Justice. Luban has been a Guggenheim fellow, a fellow of the Woodrow Wilson Center, and a fellow of the Institute for Advanced Studies at Hebrew University. He has received prizes for distinguished scholarship from the American Bar Foundation, the New York State Bar Association, the American Publishers Association, and the Centre for International Law Research and Policy.

Valerie Morkevičius is associate professor of Political Science, and Chair at the Department of Political Science, Colgate University. She is the author of *Realist Ethics: Just War Traditions as Power Politics* (2018), which uses an exploration of the history of the Christian, Islamic, and Hindu traditions to reveal that just war thinking is no stranger to pragmatic politics. Her other work focuses on the intersection between power and ethics, and the applicability of traditional just war thinking to contemporary challenges, including cyberwarfare and information warfare.

Eric Patterson is president of the Religious Freedom Institute in Washington, DC, and scholar-at-large at Regent University. He has served as a White House Fellow assigned to the U.S. Office of Personnel Management, as an Air National Guard commander for over 20 years, and twice worked at the U.S. Department of State's Bureau of Political-Military Affairs. His extensive work on war, peace, and security includes dozens of academic and popular articles and 20 books, including *Just War and Christianity: A Concise Introduction* (2023), *Just American Wars: Ethical Dilemmas in U.S. Military History* (2019), and *Ending Wars Well: Order Justice, and Conciliation in Post-Conflict* (2012).

Foreword

James Turner Johnson

Early in his chapter, "Returning Military Necessity to the *Jus in Bello,*" Eric Patterson notes the absence of treatment of military necessity by just war scholars, contrasting this with the emphasis on the concept in legal textbooks and in military doctrine and teaching. Other chapters explore in various ways how consideration of military necessity might be incorporated in moral thinking about the idea of just war. This is an important and worthwhile exploration, but one which calls for attention to an earlier question: How did these different disciplinary approaches to laying out normative limits on warfighting come to diverge on the matter of military necessity?

To push Patterson's contrast a bit further, the current situation is not simply that writers on restraint in war in the humanities and social sciences ignore the concept of military necessity but that, for at least a sizeable fraction of those who employ one or another framework built on the normative concept of just war, appeals to military necessity are understood as challenges to the restraints associated with just war reasoning, while in the legal and military frameworks, it is understood as strengthening the power of these restraints. So, the question of how just war scholarship in the humanities and social sciences came to diverge from legal and military thinking on limits to be observed in warfighting also needs to take account of this antipodal valorization of the concept of military necessity.

To speak of *returning* military necessity to the *jus in bello* implies that it once was part of the just war idea, but the real history is more complex. When a systematic conception of just war first came together in the high Middle Ages, law and morality were not conceived as separate and different endeavors but were aspects of a single whole encompassed in the idea of natural law. The earliest just war theorists were in fact lawyers: the canonists from Gratian through his two generations of successors, the Decretists and the Decretalists. Gratian's iconic work the *Decretum*, from the middle of the twelfth century, systematically defined just war in terms having to do only with resort to war. The same was true for Thomas Aquinas's treatment of just war in Part II/II, Q. 40 of his *Summa Theologiae*, which set the idea of just war in the context of theology. But the canon law did include restraints of war conduct in the forms of lists of noncombatants—persons not to be directly and intentionally attacked in war—and efforts to outlaw certain kinds of weapons, but these remained separate from the rules for resort to war, and the same was true in Aquinas's thinking. Aquinas relied centrally on Gratian both in his treatment of war and in his treatment of sexuality, and by placing both these ideas within a theological framework, he opened the door to the development of moral theology as an intellectual discipline separate from law, if not fundamentally distinct from it. Aquinas's summary of the idea of just war as a *jus in bello* requiring right authority, just cause, and right intention (avoidance of wrong intention and orientation toward good ends) opened the way for a theological definition of *jus in bello* to develop, but Aquinas himself, like Gratian, did not develop this idea. Rather, the early history of the *jus in bello* was in the frame of the rules for chivalric conduct in war. In the era of the Hundred Years' War, influential writers on chivalry and war conduct drew together the chivalric tradition and the tradition contained in the canon law, placing the *jus in bello* thus defined alongside the canonical and theological rules for resort to war. The result was a conception of just war that included both a *jus ad bellum* and a *jus in bello*. This was the understanding of just war employed by late medieval and early modern just war thinkers and known by the Dutch jurist Hugo Grotius, who used it as a base for his conception of the law of nations.

Was there a concept of military necessity in just war thinking during this early period of consolidation and definition of the just war idea? The term "military necessity" did not yet exist, but it can be argued that something of that nature was implied in the terms Aquinas identified as the requisites of resort to a just war: that is, the fundamental aims of sovereign authority, just cause, and right intention all implied not only the right to resort to armed force but also the sovereign's obligation to do so in order to rectify injustice and to seek the good of his political community (the *res publica*, or public entity; often translated "commonwealth"). Those charged with ensuring these core concerns were the sovereign ruling authorities in each political community. When all three requirements were satisfied, resort to armed force was allowed. On this reasoning, what today is called military necessity was derived from the satisfaction of the requirements of the *jus ad bellum*, and the right of using armed force carried over into how such force was to be used. This right, expressed in the conduct of the use of armed force, was limited by the justifying requisites. This was underlined in both the chivalric code and the canon law by the protection of noncombatants from direct, intended harm and by rules limiting the use of certain types of weapons, those whose use tended to cause indiscriminate harm.

This way of thinking about resort to war and right conduct in war held together only so long as the unity of the moral community of Christendom also held. When this moral community broke down as a result of the Protestant Reformation, this also affected the tradition of just war that was a product of this community. Two scholars writing early in the seventeenth century, Grotius and the Spanish Jesuit theologian Francisco Suarez, show the results of this stress on just war tradition in different ways. On the one hand, after Suarez, there were no more book-length theological or philosophical treatments of the just war idea for over three centuries, when books by theologian Paul Ramsey in the 1960s and the political philosopher Michael Walzer, then years later, began the wave of moral discussion of just war that continues into the present. On the other hand, Grotius' *On the Laws of War and Peace* refashioned the just war tradition he had inherited from the early modern just war theorists, casting the result in the form of a law of nations that depended on agreement among the political communities themselves, a system that corresponded to the international order put in place by the Peace of Westphalia that ended the Thirty Years' War, the last war of religion in European society.

The absence of theological and philosophical discussion of the just war idea that began in the seventeenth century did not mean the end of just war tradition; rather, the idea of just war continued, though with some changes, in the new context of the law of nations. A major shift here was to deemphasize the matter of the *jus ad bellum*, since the new framework of the law of nations gave the right of making war to each state. The near-term result was a century of sovereigns' wars, fought no longer over rivaling religions but rather over rivaling claims to territory by different states. In this context, the inherited tradition of just war shifted to the matter of war conduct and the *jus in bello.* Here the idea of military necessity took on greater importance, and the term itself began to be used. But in the new context, its meaning was determined by the answer to the question of what means were allowed when the right to war was allowed. That is, the idea of military necessity moved from the *jus ad bellum* to the *jus in bello.* At first, what is right conduct in war was determined by custom, and thus, custom, shaped by both the nature of the practice of war and the *jus in bello* restraints in just war tradition, determined what was understood as militarily necessary.

But the nineteenth century brought fresh challenges to the inherited consensus on right conduct in war: new, more powerful weapons of war and the move toward national mobilization and much larger armies during the Napoleonic wars. Large national armies proved to be a lasting phenomenon, and thus the effort to restrain the destructiveness of war centered on its weapons: the means of war and how they were used. In the last half of the nineteenth century, international agreements began to appear renouncing new, more destructive types of weapons or restricting their use. The first such agreement aimed specifically at a new form of weapon was the St. Petersburg Declaration of 1868 prohibiting the use in war of explosive projectiles under 400 grams in weight. This declaration also set a precedent for subsequent international law by stating in its opening lines, "the only legitimate object which States should endeavor to accomplish during war is to weaken the military forces of the enemy." This perspective was echoed in the 1899 Hague Conference Declarations, which linked the restrictions laid out there to implications in customary international law. The theme is constant within both the 1899 Hague Declarations and the 1907 Hague Conventions: what can rightly be attacked in war are military objectives, and these are always subject to attack. But otherwise, as 1907 Hague Convention IV states the matter, harm to persons and property is to be avoided because of "the desire to limit the evils of war, as far as military requirements permit." The idea of military necessity was thus linked to legitimate military objectives. This remains a fundamental principle in the development of international law on war. The other side of the matter is that military necessity does not allow the direct, intended targeting of non-military objectives. This remains a clear theme in the law of war and in literature relating to it, and it provides the basis for treatment of military necessity in the U.S. military as well. The fundamental idea has remained here that military necessity refers to the right to make war as granted by the *jus ad bellum*.

But during the same period when restrictions on the means and objectives of war began to take shape in nascent positive international law, an opposite position also developed. Because the right to make war was regarded as a right of states *qua* states, it was comparatively easy to construct an argument that justified, in the name of military necessity, anything a state might find it necessary to do in pursuing its right to make war. This argument that in a war the state may use any means necessary to secure its ends became a feature of German military thought and doctrine in the form of the idea of *Kriegsraison* (literally, the rationality of war in itself) in the late nineteenth and early twentieth centuries. *Kriegsraison* was a theory of military necessity that held that war imposed its own rules and necessities. In German thinking and doctrine about these concepts, it was contrasted with *Kriegsmanier* (literally, the manner of war), which referred to the internationally agreed upon laws of war. *Kriegsraison* trumped *Kriegsmanier*. Stark differences between the idea of *Kriegsraison* and the developing international positive law defining and imposing restrictions on the methods and means of war were fought out not only in theoretical and political debates but in war itself. In World War I, the Imperial German fleet was largely confined to the harbor at Kiel while it waited for a favorable opportunity for a sortie that would bring its main battle fleet into combat with the British fleet, based at Scapa Flow in the North Sea islands north of Scotland. While the battleship fleets waited in their respective bases, the battle cruisers on each side patrolled the English Channel, each seeking to damage the other side. In this context, the German battle cruisers frequently engaged in bombardment of shore targets, escalating from military targets (which were legitimate objectives) to civilian towns and villages, some situated close to British naval sites, others not. The German Navy cited *Kriegsraison* in defense of their bombardment of these civilian sites. The British, in contrast, denounced these bombardments as violations of the law of war. Ultimately, the matter was decided in favor of the British by their victory over the German High Seas Fleet and the overall victory of the Allies over Germany.

Ramsey's and Walzer's recovery of the just war idea in theological and philosophical moral debate did not directly address military necessity in their reconstructions of the just war idea, though Walzer took international law on war as one of his starting points. It is possible to read Walzer on supreme emergency as a treatment of military necessity—and some of the writers in this volume do—but he never used the latter term himself and subsequent developments of the military necessity idea have not built on Walzer. Both Ramsey and Walzer generated streams of just war thinking after them, and the absence of military necessity in the ensuing scholarly debates reflects their lack of engagement with this idea in their own work. I have often referred to the work of Ramsey and Walzer as two of the stools on which the late twentieth-century recovery of the just war idea rests. There was also a third stool, the United States Catholic Bishops' pastoral letter *The Challenge of Peace*, which appeared in 1983. Not only did this pastoral letter itself not take up military necessity as an element of just war thinking, but also its own reconstruction of the just war idea, shaped in response to the idea that any new war would be nuclear, held that mutual escalation would lead to a global holocaust. This way of thinking contrasted sharply with Ramsey's engagement with nuclear strategic thought and policy efforts aimed at strengthening noncombatant immunity, lowering the overall destructiveness of nuclear weapons, and reducing the likelihood of escalation to mutual assured destruction. Proponents of the bishops' position criticized Ramsey as too optimistic about the possibility of limiting nuclear war and its destructiveness. These critics, ironically, held to their own version of *Kriegsraison*, the idea that war defines its own rationality and that this presses toward greater and greater destruction so as to ensure victory. But contrary to the German General Staff in the later nineteenth and early twentieth centuries, these thinkers took their version of *Kriegsraison* as part of an argument for ending all war. Thus, ironically, the idea of military necessity became a part of the antiwar aspect of current just war thinking, where it challenges the whole historical thrust of just war reasoning aimed at accepting the use of armed force when justified by the *jus ad bellum* and channeled toward limiting harm and destruction and promoting justice and peace as provided for in the *jus in bello.*

In the present context, then, just war thinking as a moral enterprise might do well to reengage with thinking on the law of war so as to recover the common elements driving both and also to seek to make use of the law's treatment of military necessity as part of its rationale for limiting the destructiveness and harm of contemporary armed conflict. This will provide moral reflection about just war with an additional tool to seek to limit the destructiveness of contemporary armed conflict and also to oppose the idea that war in itself, because of its internal rationale pushing toward ever more destructiveness, cannot be constrained and must be morally opposed. For these reasons, this present collection, unified by a common focus but not always speaking in a singular voice, is welcomed.

1 Returning Military Necessity to the *Jus in Bello*[1]

Eric Patterson

United States and other Western militaries emphasize the limits imposed by the law of armed conflict (LOAC). In the United States, our personnel go through a LOAC refresher training annually on the *jus in bello* (ethics of how war is fought) principles, starting with military necessity; and then proportionality; and then non-combatant immunity, or distinction; and then principles of humanity and unnecessary suffering. When foreign militaries, particularly in developing countries, receive training from the United States and other Western militaries, or the excellent curriculum from the International Committee of the Red Cross (ICRC), they receive a similar experience, which emphasizes military necessity alongside proportionality, non-combatant immunity, unnecessary suffering (the principle of humanity), prohibited arms, and the like.

In stark contrast, the vast majority of books written on just war thinking in the past half century generally avoid military necessity. Among just war scholars in the humanities and social sciences, not one of our major lists of just war criteria features military necessity as a *jus in bello* principle.[2] This is in stark contrast to legal textbooks at every major law school, which emphasize military necessity.[3] In short, our lawyers and soldiers emphasize military necessity but our just war scholars largely neglect it. Indeed, a search of journals such as *Journal of Military Ethics* and *Ethics and International Affairs* shows little original scholarship on military necessity in the past two decades.[4]

This volume is designed to change that. Our purpose is to reconnect today's just war scholarship to both the rich historical tradition on necessity and contemporary strategic and legal thinking on military necessity. Although the prime area of focus in this book is on tactical and operational military necessity, i.e., what is happening on a certain battlefield at a specific time and place, the book also points in the direction of strategic forms of military necessity (e.g., campaign- and theater-level) and also considers how *jus in bello* necessity is linked to necessity within the *jus ad bellum* criteria.

This introductory chapter provides a refresher on the just war criteria and then reintroduces battlefield military necessity as a *jus in bello* principle, calling on just war scholars to reconsider this vital concept as a principle of stewardship, particularly when it comes to matters of *troop protection*, *economy of force*, *military effectiveness*, and an *aim toward victory*. As we will see in the final section, which

DOI: 10.4324/9781003390398-1

presents an overview of the experts' arguments in succeeding chapters, the return of military necessity, alongside proportionality and discrimination, provides a richer tactical and operational *jus in bello*, useful for theorists and for practitioners.

What Is Just War Theory?

The classical just war framework provides the foundation for customary international law as well as the formal laws of armed conflict, in addition to ethical reflection. Just war thinking begins with three criteria for the just decision (*jus ad bellum*) to use military force: *legitimate authority* acting on a *just cause* with *right intent*. Practical, secondary *jus ad bellum* considerations include *likelihood of success*, *proportionality of ends*, and *last resort*. Just war thinking also has criteria regarding how war is conducted (*jus in bello*): using means and tactics proportionate (*proportionality*) to battlefield objectives and which limit harm to civilians, other non-combatants, and property (*discrimination*). Historical Christian just war thinkers, such as Augustine, Aquinas, Gratian, Vitoria, Luther, and Calvin, all dealt with some form of *military necessity* and although neglected by many key theorists in recent years, nonetheless the criteria is hardwired into legal theory and military doctrine.

More specifically, just war thinking compels political authorities and their subordinates to carefully examine the following principles when making decisions about employing force:

Principles of Just War Statecraft

Jus ad Bellum

Legitimate authority: Supreme political authorities are morally responsible for the security of their constituents and therefore are obligated to make decisions about war and peace.

Just cause: Self-defense of citizens' lives, livelihoods, and way of life are typically just causes; more generally speaking, the cause is likely just if it rights a past wrong, punishes wrong-doers, or prevents further wrong.

Right intent: Political motivations are subject to ethical scrutiny; violence intended for the purpose of order, justice, and ultimate conciliation is just, whereas violence for the sake of hatred, revenge, and destruction is not just.

Likelihood of success: Political leaders should consider whether or not their action will make a difference in real-world outcomes. This principle is subject to context and judgment, because it may be appropriate to act despite a low likelihood of success (e.g., against local genocide). Conversely, it may be inappropriate to act due to low efficacy despite the compelling nature of the case.

Proportionality of ends: Does the preferred outcome justify, in terms of the cost in lives and material resources, this course of action?

Last resort: Have traditional diplomatic and other efforts been reasonably employed in order to avoid outright bloodshed?

Jus in Bello

Military Necessity: Is every reasonable effort made to gaining battlefield advantage in pursuit of larger strategic objectives, while restrained by law and other jus in bello criteria?

Proportionality: Are the battlefield tools and tactics employed proportionate to battlefield objectives?

Discrimination: Has care been taken to reasonably protect the lives and property of legitimate non-combatants?

Jus Post Bellum

Order: Beginning with existential security, a sovereign government extends its roots through the maturation of government capacity in the military (traditional security), governance (domestic politics), and international security dimensions.

Justice: Getting one's "just deserts," including consideration of individual punishment for those who violated the law of armed conflict and restitution policies for victims when appropriate.

Conciliation: Coming to terms with the past so that parties can imagine and move forward toward a shared future.

Over the past centuries, these principles have become rooted in customary international law, the war convention, international humanitarian law (IHL), and practical military ethics. The principles of sovereignty and non-intervention, for example, did not just arise from the exhaustion of the four overlapping wars that we today call the Thirty Years' War (1618–1648). When the Peace of Westphalia (1648) was brokered, the belligerents naturally fell back on the customary international principle of proper or legitimate political authority over a given territory as the bedrock concept for international society. More recently, beginning in the late nineteenth century with America's Lieber Code during the U.S. Civil War and legal developments in Europe, led in part by the then-fledgling International Committee of the Red Cross (ICRC), a legal edifice enshrining these principles has developed in treaties such as the Hague and Geneva Conventions and the Charter of the United Nations.

Military Necessity: Definitions and Controversies

As Valerie Morkevičius demonstrates in Chapter 8, the concept of strategic and operational necessity has a long tradition in the West and is also adumbrated in Islamic and Hindu texts. For our purposes, an important milestone is the explication of military necessity by Professor Francis Lieber in a set of instructions often called "The Lieber Code," issued by Lincoln's War Department as General Order 100 (1863), "Instructions for the Government of Armies of the United States in the Field." The Lieber Code is important because within a decade of the U.S. Civil War, its principles had been adopted into the military doctrine of a number of Western and Central European countries and it became the spine for what later evolved into the U.S. Code of Military Justice (UCMJ). The Lieber Code defined military necessity in Article 14 as the "measures which are indispensable for securing the ends of the war, and which are lawful according to the modern law and usages of war."[5] Later articles narrow military necessity to ensure that wanton violence, destruction, and perfidy are not labeled as "military necessity."[6] A key presupposition here is that we are talking about military necessity as tactical and operational activities on the battlefield.

Today's *U.S. Department of Defense Law of War Manual* says that military necessity may be defined as "the principle that justifies the use of all measures needed to defeat the enemy as quickly and efficiently as possible that are not prohibited by the law of war."[7] Gary D. Solis in his influential legal textbook, *The Law of Armed Conflict*, writes: "Military necessity is an attempt to realize the purpose of armed conflict, gaining military advantage while minimizing human suffering and physical destruction."[8] William V. O'Brien, in his classic text, writes,

> Legitimate military necessity consists in all measures immediately indispensable and proportionate to a legitimate military end, provided that they are not prohibited by the laws of war or the natural law, when taken on the decision of a responsible commander, subject to review.[9]

Unfortunately, military necessity is often derided, poorly defined, or inappropriately used by those claiming to take a just war perspective. Boston University scholar Neta Crawford suggests that military necessity is merely immoral consequentialism that allows military personnel to violate civilian protections.[10] Even Daniel Bell, in an otherwise excellent volume on just war thinking as Christian discipleship, calls military necessity "a trump card that overrides noncombatant immunity."[11] In both cases, the criticism is that military necessity is just a cover for extreme and destructive force. Each of the writers in this volume recognize this risk and do good work in avoiding it and calling it out wherever it is found.

One reason for this censure is an aberrant form of military necessity used by the German High Command beginning in the late nineteenth century: *Kriegsraison*. The idea here is that battlefield commanders can, and should, choose to violate customary and positive law in pursuit of battlefield victories: *Not kennt kein Gebot* ("Necessity knows of no legal limitation.").[12]

This Prussian way of thinking is problematic for two reasons. First, as James Turner Johnson, citing McDougal and Feliciano, points out, *Kriegsraison* violates existing standards, such as killing surrendered prisoners of war or targeting civilians, used by the Germans in the nineteenth century and epitomized by Germany's vicious "rape of Belgium."[13] Second, as David Luban will argue in his chapter, *Kriegsraison* wrongfully severs military necessity from its intrinsic relationship with proportionality and distinction. Just as the *jus ad bellum* principle of *just cause* does not stand alone but is intrinsically linked to *legitimate authority* and *right intention* (i.e., unlawful commands by lawful leaders are still unlawful), so too battlefield *necessity*, *proportionality*, and *discrimination* are intimately linked, informing and restraining one another.

A problem that surfaces in the just war literature is a lack of distinction between military necessity as a tactical and operational *jus in bello* concept and strategic necessity. A case in point is Michael Walzer's discussion of military necessity in a supreme emergency—the potential destruction of a political community. Against the backdrop of Israel's wars for survival in 1948, 1956, and 1967, Walzer writes; "The world of necessity is generated by a conflict between collective survival and human rights."[14] Walzer is just one of many authors who focus on the decisions that political authorities (*jus ad bellum*) make when facing strategic-level disaster, in contrast to the issues of day-to-day battlefield fighting (*jus in bello*).[15]

Another reason that military necessity has fallen out of favor has to do with the anti-military bias that has become a part of the academic discussion on the use of force, particularly among some philosophers and theologians. Particularly since the 1960s, a set of arguments have developed that make it almost impossible for any conflict to be just, either in cause or in activity. Therefore, almost every application of force is at best a lesser evil, and more likely immoral. James Turner Johnson summed up two strands of this "confusion" among theologians and among secular philosophers, in this way. First, among some Christians, particularly Catholics after the Second World War,

> confusion has emerged between the Church's commitment to its teaching on just war and what has come to be called "the Catholic peace tradition," a tradition of avoidance or renunciation of participation in armed force historically associated with the religious life but, since the Second Vatican Council, made over into a case for pacifism for Catholic laity as well.[16]

Johnson rightly notes that this confusion is exacerbated by a novel idea that just war thinking is characterized as a "presumption against the use of force" rather than its historical standard of presuming that political authorities must protect order, defend the vulnerable, and advance justice.

Johnson goes on to write of a second confused strand of just war thinking:

> On the other hand, a line of interpretation has developed that has been influenced by the secular philosophical concept of prima facie duties, by prudential (and contingent) judgments about the inherent immorality of contemporary war, and by well-intentioned but rather utopian investment in the United Nations system.[17]

The kind of revisionist literature Johnson is referring to hamstrings public officials to almost never employing force without sanction from the UN Security Council, even in times of dire emergency. Even worse, this literature has developed ingenious, but deeply flawed, arguments that so limit the activities of commanders that they dare not act without the written approval of lawyers, restricting soldiers on the battlefield so that they are in far greater danger than necessary.

This book provides a needed counterpoise to such arguments. Future chapters will demonstrate the moral and prudential utility of the military necessity principle.

This raises a further question: Where should we locate military necessity among the criteria? Does it come first, second, or third on the *jus in bello* list? Does order matter? Such debates do matter among just war scholars, a case in point being whether *just cause* or *legitimate authority* has pride of place among *jus ad bellum* criteria. If *just cause* comes first, it suggests that private and vigilante "rough justice" trumps government authority. Scholars from Aquinas through James Turner Johnson and Eric Patterson argue, in contrast, that the just war tradition is rooted in civilizational principles of the rule of law and the responsibility of authorities within their given sphere of influence. The use of force should rightfully be in the hands of proper authorities who have the responsibility to protect and defend. If there are to be exceptions, then those must be fleshed out as the rare exceptions that prove the rule.

So, too, it does matter how we think about the principle of military necessity within the *jus in bello* categories. Some scholars, such as Nigel Biggar and Keith Pavlischek, suggest that if we reintroduce military necessity, we should consider it a part of the principle of proportionality.[18] Neta Crawford agrees, saying that proportionality implies an evaluation of military necessity.[19]

Others suggest that military necessity, if a part of the *jus in bello* criteria, is at all times in tension with, or restrained by overriding principles of proportionality and discrimination. Hence, they would suggest that the ordering of the *jus in bello* criteria should be proportionality first, then discrimination, and then military necessity.[20]

We argue that it is important to put military necessity first, as in the law of armed conflict literature, while recognizing that *military necessity* is in a working relationship with *proportionality* and *discrimination*. It is best not to see the *jus in bello* criteria as a mere checklist, but rather see all three principles in an overlapping, interdependent relationship to one another. This is true for all just war principles: they are an interdependent whole providing guidance for moral decision-making by authorities who are responsible for action.

Military Necessity as Stewardship

Military necessity is concerned with the practical aspects of battlefield scenarios that must be taken into account in light of the imperatives of troop protection and victory.[21] Military necessity should be particularly important to democracies because it protects the lives of citizens in uniform. More specifically, military necessity is a national security stewardship principle for reasons of *troop protection*, *economy of force*, the aim of *victory*, and *military effectiveness*.

Military necessity as stewardship is the careful management of that which has been entrusted. What is a greater trust than human life and property? Military necessity does include cost-benefit analysis, but this is not craven consequentialism. With its roots in Christian just war thinking, stewardship is extremely important. There are many biblical passages about kings and princes counting the cost before building defenses or taking action, and such remains true for political authorities to this day. From a moral philosophy and democratic theory perspective, leaders are to be responsible, and held accountable, as custodians of the lives, livelihoods, and way of life of their fellow citizens. Military necessity makes investments in their safety part of the calculus of how battles are to be fought and won.

There are at least four elements of military necessity that are important for national security stewardship. The first is *troop protection*. Military commanders should be deeply concerned with protecting the lives and welfare of their own troops and this should influence the way they make decisions about battlefield activity. Moreover, military personnel are citizens who, when at home, are husbands and wives, sisters and brothers, children and parents. Soldiers are citizens, whether conscripted or voluntary, and we should safeguard their lives in accordance with the other *jus in bello* principles of proportionality and discrimination.

The second element is the principle of *economy of force*. Traditional military strategy has a set of principles such as not dividing one's forces, focusing one's forces in a certain area, and only using as much force as needed in a given domain. These are prudential considerations that have to do with husbanding one's resources. Such resources are not finite and are funded by the tax dollars of average citizens. These resources include more than military materiel, most importantly the lives of men and women in uniform. Moreover, when force is used in a deliberate but economical way, this means far less destruction of the enemy, far less rebuilding, and far less collateral damage.

Third, military necessity is also about *effectiveness*. Effectiveness means how well the job gets done. Historians are rightly concerned about how sloppy planning and ineffective battlefield plans occasioned massive destruction in the First World War, such as at Somme. In contrast, we laud the effectiveness of the battleplan in the 1991 Gulf War. Effectiveness is an important part of military necessity: Will this plan work? Will our actions be effective? Effective actions are far less wasteful and destructive and this is a feature of moral accountability.

Fourth, military necessity is a link between battlefield activities and stated war aims (*jus ad bellum*) as well as a link between what happens now and the long-term accomplishment of war aims in *jus post bellum*. In other words, military necessity has an eye at all times on *victory:* victory in this skirmish, victory on this battlefield, victory in this campaign, and, ultimately, how this battlefield engagement advances long-term strategic victory. Too often arm-chair theorists want to use *jus in bello* to limit the activities and armaments of troops at the site of a specific tactical encounter. Military necessity reminds us that each battlefield is linked to larger campaigns in the grand strategy of the conflict.[22]

In short, military necessity is a critical stewardship principle that includes principles of *troop protection, economy of force, military effectiveness,* and an *aim toward victory* that is important in providing a richer *jus in bello* that accompanies the principles of proportionality and discrimination. It is a moral criterion long recognized in the law of armed conflict and in legal treatises, but that has fallen out of favor in recent years with just war scholars in the social sciences and humanities. It is time to return military necessity to *jus in bello* just war thinking.

Overview of the Book

The first chapters of this volume look at military necessity conceptually, through its historical development and how it has been deployed in natural law reasoning, classical and Christian thought, and in light of contemporary military ethics. In Chapter 2, J. Daryl Charles argues that too often the principle of military necessity has been derided as *realpolitik* or some form of amoral pragmatism. But, when one begins with the obligations found within natural law reasoning we can see a deeper moral ethic for military necessity. One way of describing the natural law's presence and universality is to understand it in terms of the "Golden Rule" described by Plato, Jesus, and others. *No one* – anywhere on the planet – does *not* believe the "Golden Rule." And what *is* that rule, that moral norm? Positively, it is to act and treat others as we ourselves would wish to be treated. Negatively, it is not to act and treat others as we ourselves would not wish to be treated. Yet a third (and negative) corollary issues out of the "Golden Rule" ethic, and this has enormous ramifications for both humanitarian and coercive military intervention: namely, we must not *allow others* to be treated as we ourselves would not wish to be treated. Protecting the neighbor expresses both justice and charity; this *is* the "Golden Rule," which guards human dignity. This chapter lays a natural law foundation for the *jus in bello* criteria with a particular focus on how *right intention* informs the interdependent trio of *military necessity*, *proportionality*, and *discrimination*.

Christian Braun builds on these themes in Chapter 3 with a careful analysis of the Thomistic and Medieval roots of military necessity and how they inform today's debates. Beginning with Thomas Aquinas's emphasis on training in virtue, Braun engages with the military necessity principle as found in the classical *bellum justum*. He points out that military necessity in the above understanding was an important consideration in the Middle Ages and discusses the divergent attempts to regulate war conduct advocated in canon law and the code of chivalry. Braun then turns to Aquinas's understanding of virtue and explains why he gave great emphasis to virtuous behavior in war and did not put forward a detailed *jus in bello* code. Building on the crucial place for virtue in Aquinas's just war, Braun turns, in the conclusion, to the debate about the right place of military necessity (i.e., the debate between those like Patterson who call for its return and those who neglect it like Walzer and the revisionists) in today's just war *jus in bello*. I will argue that the main contribution the Thomistic just war can make in this regard is to create virtuous soldiers that will naturally be able to take the right decisions in the heat of battle and, therefore, apply military necessity correctly.

In a revised and updated version of a classic paper, David Luban in Chapter 4 explores the way that military necessity functions in moral and legal discourse about the *jus in bello*. Guiding the reader on a lexigraphical tour of the various legal definitions of military necessity, Luban illustrates how these are not solely a matter of positive legal and doctrinal arguments but also reify moral definitions. Luban ultimately lands on a defense of the "marginal" view of necessity. Behind this concept is the idea that the true military significance of an act is the gain it provides over the next-best alternative; it is "necessary" only in the sense that it is necessary to achieve this marginal gain in military advantage.[23] Necessity judgments are inherently comparative. If the advantage over less harmful alternatives is too small, the claim of military necessity lacks normative force.

In Chapter 5, Louis Bujnoch adds to our understanding by examining the different ways that military necessity is conceptualized and implemented by different types of authority. More specifically, the concept of necessity is frequently invoked in relation to war and conflict in a variety of contexts ranging from specific legal provisions (e.g., narrow military necessity in the Law of Armed Conflict) to broad political-strategic claims in reason of state logic. The latter have found expression from the morally flawed *Kriegsraison* of the nineteenth century to Henry Kissinger's modified "flexible response" in his 1960 book *The Necessity for Choice* to various iterations of the U.S. National Security Strategy. The conceptualization of necessity as well as the purpose and intended effect of an invocation thereof, are substantially varied and often at odds with one another. Bujnoch's chapter examines this broad range of conceptualizations and proposes a dichotomous heuristic centered on the dual meaning of necessity as both inevitable and indispensable to cut across the cacophony of conceptualizations of necessity.

As noted above, Catholic thinking on issues of war and peace has been particularly contentious since the Second World War. In Chapter 6, Pedro Erik Carneiro helps us understand change and continuity in Catholic thought. After introducing some of the crucial historic Catholic just war thinkers who laid the foundation for thinking about military necessity, Carneiro focuses on modern Catholic thinkers who, since the twentieth century, have debated issues of military necessity. In particular, he looks at some of the Catholic thinkers who defended the extreme necessity (*kraigsraison*) associated with Prussian military and, later, with the Nazis. Other Catholic thinkers, such as Elizabeth Anscombe, were critical of those forms of necessity and, in particular, how the idea of necessity was used to defend Allied fire-bombing of Axis cities and, ultimately, the use of atomic weapons. Catholic discussion and debate about military necessity continued through the Cold War, with topics ranging from national defense to national liberation. These debates also help us understand the linkage of *jus in bello* and *jus ad bellum* conceptions of necessity. As the Catholic Church continues to be a critical source of just war teaching for religious, academic, and military audiences, the evolution of these debates by key figures, including at least two popes, is useful for reinvigorating discussion on military necessity today.

In Chapter 7, the U.S. Naval War College's Pauline Shanks Kaurin asks whether and how military necessity ought to have status as a distinct moral requirement within the context of just war thinking. Accordingly, this chapter proceeds in three sections. First, a discussion on the nature of military necessity followed, second, by an examination of military necessity as a separate just war criterion. Kaurin then proceeds to some considerations for moving the conversation forward including an argument that any military necessity requirement must be conceptually distinct from proportionality judgments, and that judgments about military necessity must not merely focus on military culture, preference, and procedures, but also consider the impact of those harms from their point of view using methodologies appropriate to their experiences and concerns.

Building on her work about "realist ethics," Valerie Morkevičius in Chapter 8 asserts that just war thinking invokes necessity in two ways: not only as a restraint (as when arguing that a particular war is unnecessary) but also as a permission, or even a duty (as when "the failure to go to war cannot but result in injustice").[24] Her chapter aims to disambiguate these two profoundly different ways in which necessity functions as a principle. While necessity restricts the scope of the just cause principle and underlies the principle of last resort, necessity is also used to explain why the use of force is ever justified in human affairs and to loosen some of the *jus in bello* principles that strategists and war-fighters might otherwise find too idealistic. To this end, this chapter explores each of these facets of necessity through the lens of the historical just war traditions as they evolved within Christianity, Islam, and Hinduism.

Shannon French has spent many years training future military leaders. Collaborating here with her husband, J.A. French Flint, in Chapter 9 they consider military necessity in its dimensions as both a constraint preventing certain kinds of actions as well as a goad *demanding* certain kinds of behavior. In contrast to Morkevičius's approach from realist ethics, French and French Flint focus on the responsibilities and consequences raised by military necessity. For example, where military necessity acts as a goad—essentially saying "This thing must be done"—there arises a possible tension. This "thing" which "must be done" might be so heinous that it is likely to lead to various forms of moral trauma, including moral injury. With this rather obvious possibility in view, this chapter then proceeds to consider what it might mean for officers to be held responsible for safeguarding not just the lives of their troops, but also the humanity of their troops. How should such a charge be understood, and can it be justified? If officers are to be held responsible for protecting their troops in any way beyond the physical, it must be against specific, severe threats to their humanity that occur in the course of waging war. Candidates for threats of this kind are considered, leading to the conclusion that the greatest threats arise from *jus in bello* violations that dehumanize the victim and degrade the perpetrator. The question is then raised whether officers in fact can protect their troops from committing such violations, and the argument is advanced that the command climate that officers create in their units plays a significant role in encouraging or deterring serious transgressions of the warrior's code.

Amos Guiora in Chapter 10, turns our attention to the aftermath of Munich 1972. The context is the massacre of Israeli athletes by Palestinian terrorists, a situation dramatically heightened by the fecklessness of the German security services and the fact that the slaughter occurred on German soil less than 30 years after the Holocaust. The response, ordered by Prime Minister Golda Meir, was conducted by the Mossad, the Israeli national intelligence service that operates outside of Israel's borders. There was no discussion of capturing-detaining-interrogating the identified Black September operatives; the operational goal was clear and concise, devoid of nuance: kill those deemed responsible for the murder of the 11 Israelis. The murders were conducted in a variety of ways, including shooting, bombs placed in telephones, and under an individual's bed; the methods were intended both to kill the intended target and to send a clear message to others. The operation was halted after the mistaken-identity killing of a Moroccan waiter in Lillehammer, Norway.

The common consensus in the operation was revenge-based, which is not tolerated by international law. That is distinct from self-defense, a right clearly granted to governments in Article 51 of the United Nations (UN) Charter. While the UN Charter defines self-defense in the context of state-state conflict, it is applied by nation states engaged in operational counterterrorism when engaged with terrorist organizations (non-state actors). Prime Minster Meir gave short shrift to questions of morality and self-imposed limits of state power. There is no suggestion this was a relevant concern from her perspective; rather, her focus was exacting (word used deliberately): "a price for the attack." Consequently, in assessing whether the Operation met standards of morality, such as *just cause*, *discrimination*, and *military necessity*, we need to examine the question from two distinct perspectives: Prime Minister Meir "in the moment" and retrospectively with the benefit of time and distance. To understand the Israeli public mood of 1972 requires recognizing that the trauma of the Holocaust was very much "front and center," not to mention that Israel had just recently survived the third of a series of wars with its neighbors.

In Chapter 11, Regent University's Joshua Hastey takes us to one of the least understood yet important battlefields of our time, what scholars call the Hobbesian "gray zone" between peace and all-out war. For more than a decade, scholars and warriors have considered this arena, which ranges from "military operations other than war" to cyber-security, information operations, battle preparation and war prevention, espionage, surveillance, and other activities. Hastey argues that military necessity has typically only been evaluated in times of large-scale wars in progress. This framing, consistent with the just war tradition's concern with limiting the harm inflicted by war, sets the doctrine in tension with the Clausewitzian emphasis on economy of force, with the former advocating restraint for the moral purpose of harm aversion and the latter counseling local restraint to permit the application of greater force at the enemy's center of gravity. His chapter presents an alternative, complementary framing of military necessity as the natural extension of the *jus ad bellum* standard of right intent, and allows the long-neglected military necessity to provide crucial moral clarity in the fog and friction that pervade conflict in the gray zone.

Offering a case examination, Marc LiVecche in Chapter 12 takes us to one of the most highly debated strategic decisions of the past century, President Truman's decision to employ the atomic bomb against Japan. LiVecche's chapter proceeds from the just war recognition that simply because something is deemed necessary it does not necessarily mean that it is moral, as when it violates *jus in bello* commitments to proportionality and discrimination. Further, LiVecche takes for granted that what is not moral is therefore not permitted. At the same time, this chapter argues that in certain circumstances, necessity carries its own moral weight, guiding our assessment of both proportionality and discrimination. To get at these issues, his chapter explores the case of the dropping of the atomic bomb on Hiroshima and Nagasaki, Japan. He argues that given the context in the Pacific Theater from the summer of 1944 to August 1945, using the atomic bomb was both necessary and moral. In fact, it is precisely because the bomb was necessary to achieve victory through an unconditional surrender in the shortest possible time that it was moral and in keeping with just war principles.

Finally, in a concluding chapter, I offer a summary overview of military necessity as an essential element of national security stewardship. Canvassing four components of stewardship—force protection, economy of force, effectiveness, and victory—I link necessity to the principle of *jus post bellum* and examine why the linkage is important. Bringing the chapter to a close, I point out several critical areas that are ripe for further research as we continue to explore the vitality and usefulness of the idea of military necessity for just war scholarship and statecraft moving forward.

Notes

1 I thank Keith Pavlischek and James Turner Johnson for their insights on this argument. The original concepts were presented at a just war working group hosted by Nigel Biggar and Marc Livecche at Oxford University and then at a later working group in Washington, DC, both supported by the McDonald Agape Foundation. I am grateful to those at the working groups who provided helpful criticism, including Daniel Strand, Debra Erickson, John Kelsay, Joseph Capizzi, J. Daryl Charles, Josh Hastey, Chris Eberly, Rich Frank, and Mark Mattox. This work has had support from my research assistants, Abigail Lindner, Grace Lee Par, and Linda Waits Kamau of Regent University.
2 This survey includes my own work, which provides the standard narrow list of *jus in bello* criteria: discrimination and proportionality. The one important outlier was pointed out to me by James Turner Johnson, who referred me to William V. O'Brien's exceptional *The Conduct of Just and Limited War* (New York: Praeger, 1981). O'Brien develops military necessity at length, relying in part on expert analysis of military documents.
3 Textbook examples include Gary Solis, *The Law of Armed Conflict: International Humanitarian Law in War* (Cambridge: Cambridge University Press, 2010); Geoffrey S. Corn, et al., *The Law of Armed Conflict: An Operational Approach* (Aspen Casebook) (London: Aspen Publishers, 2012).
    Separately, there is a large set of more specialized books by legal scholars on these topics, including Yishai Beer, *Military Professionalism and Humanitarian Law* (Oxford: Oxford University Press, 2018), which begins with a chapter entitled "Revitalizing Military Necessity," and Jens David Ohlin and Larry May's, *Necessity in International Law* (Oxford: Oxford University Press, 2016).
4 A notable exception is the thought-provoking article from a 2005 edition of *Journal of Military Ethics*. See Asa Kasher and Amos Yadlin's "Military Ethics of Fighting Terror: An Israeli Perspective" (vol. 4, no. 1) and helpful responses, including those by David Perry and Nick Fotion.
5 Richard Shelly Hartigan, *Lieber's Code and the Law of War* (Chicago: Precedent Press, 1983). The eponymous Francis Lieber wrote the code as part of a five-person committee set up under General-of-the-Armies Henry W. Halleck to amend the Articles of War. That board was chaired by General Ethan Allen Hitchcock.
6 Lieber's code has three articles dealing with military necessity (Articles 14–16).
7 U.S. Department of Defense Law of War Manual, 2016, 52. Available at: https://dod.defense.gov/Portals/1/Documents/pubs/DoD%20Law%20of%20War%20Manual%20-%20June%202015%20Updated%20Dec%202016.pdf?ver=2016-12-13-172036-190.
8 Gary Solis, *The Law of Armed Conflict: International Humanitarian Law in War* (Cambridge: Cambridge University Press, 2010).
9 William V. O'Brien, *The Conduct of Just and Limited War* (New York: Praeger, 1981), 9.

10 Neta C. Crawford, "Bugsplat: U.S. Standing Rules of Engagement, International Humanitarian Law, Military Necessity, and Noncombatant Immunity," in Anthony F. Lang Jr., Cian O'Driscoll, and John Williams, eds., *Just War: Authority, Tradition, and Practice* (Georgetown: Georgetown University Press, 2013), 231–250.

11 Dan Bell, *Just War as Christian Discipleship: Recentering the Tradition in the Church Rather Than the State* (Ada, MI: Brazos Press, 2009), 22.

12 Scott Horton, "Kriegsraison or Military Necessity?," *Fordham International Law Journal* 30, no. 3 (2006): 585.

13 James Turner Johnson, *Just War Tradition and the Restraint of War: A Moral and Historical Inquiry* (Princeton, NJ: Princeton University Press, 1981), 90–91.

14 Michael Walzer, *Just and Unjust Wars*, third ed. (New York: Basic Books, 2010), 251–252.

15 In a helpful book, Michael Gross uses the principles of military necessity, effectiveness, and proportionality and talks about a two-step distillation process for making decisions about battlefield necessity, first by deciding what is necessary and then second, taking a clear-eyed look through the lens of humanitarian principles on such actions. Gross goes on to suggest an evaluation or feedback for new weapons and tactics, considering military necessity, limits of humanitarianism, and then re-evaluating the effectiveness of necessity and the limitations: Did it work as expected? Do we need new regulations or changes as we go forward? Clearly this is intelligent and ethical, but this reflective approach is the purview of military and civilian leaders, not lieutenants and soldiers under fire. See Michael Gross, *Soft War: The Ethics of Unarmed Conflict* (Cambridge: Cambridge University Press, 2017).

16 James Turner Johnson, "Just War, as It Was and Is," *First Things* 149 (January 2005): 19.

17 Ibid.

18 These points were made in a conversation with the author.

19 Neta C. Crawford, "Bugsplat: U.S. Standing Rules of Engagement, International Humanitarian Law, Military Necessity, and Noncombatant Immunity," in Anthony F. Lang Jr., Cian O'Driscoll, and John Williams, eds., *Just War: Authority, Tradition, and Practice* (Washington, DC: Georgetown University Press, 2013), 231–250.

20 I am deliberately avoiding the debate about whether proportionality or discrimination has pride of place.

21 Eric Patterson, *Just War Thinking* (Lanhan, MD: Lexington Books, 2007).

22 James Dubik makes a similar argument in the introduction to his *Just War Reconsidered* (Louisville, KY: University of Kentucky Press, 2016). He argues that we need to focus more attention on the "gap" between the *jus in bello* direction that senior leaders must provide and the *jus in bello* activities going on at the tactical level; and he proposes that the former needs more attention.

23 Seth Lazar, "Necessity in Self-Defence and War," *Philosophy & Public Affairs* 40(1) (2012): 3–44.

24 Stanley Hauerwas, "Should War Be Eliminated? A Thought Experiment (1984)," in John Berkman and Michael Cartwright, eds., *The Hauerwas Reader* (Durham, NC: Duke University Press, 2001), 417.

2 Natural Law and the Just War Ethic

Reaffirming Common Moral Traditions

J. Daryl Charles

Classic just war moral reasoning does not deny or reject the possibility of war or coercive intervention. Rather, it measures its necessity and its prosecution on the basis of moral criteria that are identified by *ad bellum* and *in bello* categories. These criteria – for example, right intention, military necessity, discrimination, and proportionality – are interlocking. They constitute a unity. A moral symmetry exists between ends and means, between a course of action and the aim that is intended. Any and all considerations informing whether or not to go to war (*jus ad bellum*) with how to conduct war (*jus in bello*) issue out of the same moral framework, and this framework is established by natural law moral reasoning – i.e., do good (justice) and avoid doing evil (injustice).[1] War or coercive force is justifiable if and where it is directed by morally good intentions and actions. Given the relative inattention to the linkage between natural law and just war moral reasoning (even among just war advocates),[2] examining natural law as a moral framework is thus our first order of business.

The Nature and Efficacy of Natural Law Thinking

Natural law moral reasoning proceeds on the baseline assumption of a shared nature in all human beings. That nature flows from two moral realities: human design based on the *imago Dei* – which demonstrates itself universally in human dignity and worth as well as the ability to reason and reflect – and human limitation – by which all human beings fall short not only of moral perfection but of their own moral standards. In characteristically Thomist language, we may say that humans are both the crown jewel of creation and the scum of the earth.[3]

Central to a shared human "nature" is a moral intuition that discerns between good and evil, justice and injustice, and expresses itself at the most rudimentary level by eschewing evil and doing good (so Thomas Aquinas).[4] It is "natural" insofar as it bears witness to our fundamental nature and design; it is a "law" to the extent that it requires of human beings to act meaningfully in accordance with their design. Thus, natural law provides a common moral grammar for all people – for people of religious or nonreligious persuasion – and facilitates moral discourse in diverse social and cultural contexts.

DOI: 10.4324/9781003390398-2

It needs emphasizing that human beings demonstrate this shared moral intuition – we may call it a "moral sense" – on a daily basis in the most basic of ways, our theoretical attempts at denial notwithstanding. Most of our daily decisions are not fused with moral significance. In traveling, we have the option of going by car, bus, train, or airplane. And while traveling, we might opt for yogurt, a hamburger, a burrito, or Chinese food. If, however, we choose to eat human flesh rather than hamburger, and if the purpose for our travel is to eliminate human beings in the service of the Mafia, then both our diet and our travel acquire a moral cast and undermine the common good. One need not be religious to intuit the difference, and thus, to acknowledge the reality of good and evil. Based on the natural law, some human actions are forbidden and some are required, given our essential nature and our common humanity. In fact, as Hugo Grotius observed, these moral-legal restraints and duties would apply even if there were no God.[5]

And yet despite the natural law's "self-evident" nature (hence, the language of the American founders), the fact of natural law thinking tends to go against the mainstream of modern and contemporary thinking. This is true both inside and outside of the academy. One obstacle is how to interpret human "nature." Are we selfish (so Hobbes and the Christian moral tradition)? Then perhaps natural aggression is morally right and normative (so Hobbes and Machiavelli). Are we capable of doing what is right and just (so Aquinas, following the Christian moral tradition)? Then despite the human inclination toward selfishness and destructive tendencies, human beings still are capable of doing what is right. Is our nature implanted by "biology" (so the materialist) or by the Creator? If the former, then evolutionary theory and "brain science" are the keys to morality, with the result that humans are *not* morally accountable for their actions. If, by contrast, nature is to be understood in terms of how we are "wired" based on the Creator's likeness and design, mirroring purpose and intent, then human beings not only can reason and reflect, but they can also intend to act, in the awareness that they are morally accountable for their actions. Based on the evidence (or lack thereof), we may conclude that it is doubtful whether moral principles forbidding particular types of actions or behavior emanate from biology. And, based on the evidence within species, we may reasonably conclude that humans and animals are fundamentally distinct, not only in terms of their dignity but also their implanted ability to reason and reflect, whereby they then act on those reasons and reflections.

Other obstacles to affirming natural law thinking in the present day arise, and they extend beyond biology per se. The influence of utilitarianism, legal positivism, developments in the social sciences, international law, certain trends within capitalist economics, and indeed a deficient theological framework contribute to suspicion toward – if not rejection of – natural law thinking.[6] However, because of its metaphysical nature, natural law must be understood philosophically, for it addresses epistemological, anthropological, ethical, theological, and – yes – political realities.

Classically speaking, the just war tradition has always had a moral-theoretical foundation, by which the "cardinal" virtue of justice and the "theological" virtue of charity operate in symbiosis.[7] Within natural law moral reasoning, wars are justifiable if they are directed by morally good intentions and actions. War is not some otherworldly realm that has little or nothing to do with how we lead our lives. Moral reflection lies at the heart of natural law thinking; that very moral reflection should prompt us to reflect on whether or not to undertake war. Undertaking coercive intervention should only be for the purposes of establishing – or restoring – justice and a *justly-ordered peace*.[8] After all, thieves, pirates, and terrorists do their best to maintain an orbit of "peace" in order that they can continue their unjust operations.[9] A "just war," then, cannot intend anything other than fostering a just peace, the common good, and human flourishing.

Contemporary resistance toward natural law thinking from various corners, it needs emphasizing, says *nothing* about its efficacy or its viability. If our contemporaries insist "Who's to say that eating human flesh or murdering innocent people for the Mafia is wrong?" the cultural relativist – we can be sure – will complain when he or she is the victim. As it turns out, Aquinas was correct: natural law thinking is a very *practical* and not speculative science. It demonstrates itself by human actions and human reactions.[10] It points to what T.S. Eliot called the "permanent things," the very moral realities that inhere in the entire created order. The natural law, then, is the foundation for justice,[11] bridging ethics and law/politics as well as moral principle and international law.

What unites all people is a general knowledge of moral first things – think, for example, of the Ten Commandments, a sort of compendium of the natural law. This "prior" knowledge or intuition – what one social philosopher calls "what we can't not know"[12] – accords with what the "apostle to the Gentiles" in the New Testament describes as the law "written on the heart." As evidence of its presence, pagan Gentiles are said to "do by nature the things required" by this law. Thereby "they show the [law's] requirements" to be "written on their hearts." For this reason, then, their "consciences [are] bearing witness" to this moral law.[13] This is none other than the language of the natural law. It is "nature" because it describes *the way things are*; the fact that not everyone gives assent to the natural law does not assail its universal reality. And it is "law" because *nothing can change our essential nature*.[14] For this reason, justice is non-fluid; and for this reason, wars can be "just" or "unjust."

Another way of describing the natural law's presence and universality is to understand it in terms of the "Golden Rule" described by Plato, Jesus, and others. *No one* – anywhere on the planet – does *not* believe the "Golden Rule."[15] And what *is* that rule, that moral norm? Positively, it is to act and treat others as we ourselves would wish to be treated. Negatively, it is not to act and treat others as we ourselves would not wish to be treated. Yet a third (and negative) corollary issues out of the "Golden Rule" ethic, and this has enormous ramifications for both humanitarian and coercive military intervention: namely, we must not *allow others* to be treated as we ourselves would not wish to be treated.[16] Protecting the neighbor expresses both justice and charity; this *is* the "Golden Rule," which guards human dignity. Both Martin Luther and John Calvin believed that the "Golden Rule" was simply a restatement of a higher inviolable law or norm, anchoring a moral universe, by which human deeds are judged.[17] If it is true that there exist moral norms and imperatives – for example, "Do unto others ..." and "Do not do unto others ..." – that transcend race, culture, ethnicity, and country, then we have moral standards and principles by which to judge human actions, including military actions and the profession of arms.[18]

Natural Law Thinking and Just War Moral Reasoning

As James Turner Johnson well observes, the challenge for any cultural apologist is to find the proper language, terms, and grammar to be employed, hence the importance of a mode of reasoning and language that serve as a "bridge."[19] In our day, of course, the secularist will be dismissive of "religious principles." The main problem, then, is lodged at the interchange of faith commitments and broader society. A natural law grammar, thus, has the utility of providing not conversion but *persuasion* in the public sphere. How, in meaningful and morally responsible ways, do we relate to those around us moral reality and moral truths, which transcend any cultural, socio-political or military realities that confront human beings?[20]

Early-modern just war theorists, located in a period of "transition" from the late-medieval to the early-modern era (from "sacred" to "secular," as it were), are significant because of their "bridge" function. Inter alia they provide an important adaptation of just war principles, which had been refined in the Christian moral tradition, to non-Christian societies and cultures. Three individuals in particular are worthy of note: Francis de Vitoria (148–1546), Francisco Suárez (1548–1617), and Hugo Grotius (1583–1645), the latter considered by many to be the father of international law. Vitoria, Suárez, and Grotius understand justice to have deeper roots than mere religious confession. Justice is known through nature and is intuited universally as binding upon all people everywhere. Therefore, the "law of nature" becomes a "law to the nations" (*jus gentium*),[21] holding people-groups and nations accountable to the unchanging demands of justice, which, in turn, orders right relations and undergirds a true and ordered peace. Just war principles, then, find confirmation in the natural law and not solely an appeal to religious faith.[22]

By about 1510, disquieting reports had been reaching Spain that Native Americans were being denied basic liberty and property. The immediate challenge confronting Vitoria was the Spanish vision to colonize the peoples of the new world. Vitoria's task was to challenge the Spanish king on the basis of unjust treatment of the American Indians. His argument was nothing short of scandalous: neither the king nor the pope for that matter could authorize war against the Indians. The only just cause for war was a wrong that was intuited through natural moral law, a wrong that is discernible to all people everywhere through reason. In Vitoria, we find the beginnings of international law, that is, of principles anchored in natural law moral reasoning that govern all nations.

Central to the work of Suárez was an emphasis on natural law and the matter of how states are to conduct themselves. Whereas civil law is alterable, the law of nature is universal and unchanging, governing how human beings and nations deal with one another. All aspects of justice flow from this reality. Under the rubric of charity, Suárez scrutinized the role of the state in both defensive and offensive modes: "A required mode and uniformity as to it [warfare] must be observed at its beginning, during its prosecution and after victory."[23] This mode of moral conformity, Suárez was careful to maintain, is founded upon the natural law and is common to religious and nonreligious people alike.[24]

Considered the father of modern international law, Grotius confronted the dilemma of just limits to war in much the same way as Vitoria and Suárez. The results of his efforts would be foundational for just war thinking in the modern era. In his important work *De Jure Belli ac Pacis* (*The Law of War and Peace*), published in 1625, Grotius argues that how nations relate to another is governed by universally binding moral principles. These are "binding on all kings" (1.1.10) and "known through reason (1.13.16). The implication is that this places limitations on whether nations may go to war and how warfare is to be conducted. Given the reality of the natural law, such rules of engagement, he insisted, are valid for all people. Natural law moral reasoning suggests neither that *nothing* is permitted nor that *everything* is.[25]

Natural law thinking presupposes the existence of universal moral norms as well as a basic awareness of these in all people. For this reason, natural law thinking is foundational to just war moral reasoning. The very premise on which "just war" rests is that there is a universal "moral sense" that informs human beings concerning their intentions and their actions.[26] It informs them, in relative terms, as to what is just and good over against what is unjust and evil.[27] Something is forbidden because it is wrong *in sic*. Of course, the question of *how we discern and decide*, procedurally and politically in terms of policy, is a secondary – though by no means unimportant – matter, calling for wisdom and prudence.

The Unity and Integrity of the Just War Ethic

Natural law moral reasoning causes us to conform to moral reality, to the principles of "right reason" and not mere legal casuistry. In its application, the natural law tradition knows no division or separation between *ad bellum*, *in bello*, and *post bellum* considerations. All three phases of just war deliberation are organically unified. The need to re-state this basic assumption is considerable, given how often in contemporary "just war" debate and discussion the families of moral criteria are viewed as independent of one another. Against many modern voices on the subject, classic just war thinking assumes an organic unity; natural law moral reasoning requires such.

In bello conditions, thus, are grounded in *ad bellum* considerations and not independent of or in isolation from them. *Jus in bello*, rather, arises from satisfying *ad bellum* conditions in general and the criterion of right intention in particular. A fundamental error of those who advance the "independent" thesis, as Anthony Coates has succinctly argued, is that it views permission and restraint as opposites rather than as a unity: "The just conduct of war *is* the restrained conduct of war."[28] The means, we might say with Hugo Grotius, are *the end in the making*.[29]

Complicating the argument of unity of just war categories is the matter of how we define 'military necessity.'[30] In his examination of military necessity, Michael Walzer has perhaps most influentially argued that 'necessity' can be overridden in the case of a "supreme emergency." In its essence, Walzer's position is that military necessity can overrule moral criteria in extreme situations.[31] Walzer writes:

"Utilitarian calculus can force us to violate the rules of war only when we are face-to-face not merely with defeat but with defeat likely to bring disaster to a political community."[32] But just war moral reasoning resists the pull toward consequentialism. War, as classic just war thinking understands it, is always an instrument of law; hence, moral constraints and limited means must always apply.[33] Not what *may* be done but what *should* be done and *how* are the burdens of just war moral reasoning.[34] 'Military necessity,' therefore, needs severe qualification in terms of definition; it is to be understood as operating *within* the constraints of the just war tradition, not apart from it. We may thus define 'military necessity' in terms of both *permission* and *limitation.* It proceeds in accordance with morally legitimate and just ends and operates within the *in bello* parameters of discrimination and proportionality.[35] Military necessity, which must be subordinated to natural law moral conditions, is expressed through and measured by proportionality – that is to say, the appropriate 'economy of force.' No more and no less; only that force which is necessary for military success is morally justifiable.[36] In this regard, while the expression 'economy of force' might suggest utilitarian measurements, within just war moral reasoning it suggests both upper and lower limits, which is to say, the *minimum* rather the *maximum* force necessary to achieve just ends.

To divorce *ad bellum* and *in bello* considerations is to forsake moral judgment and to succumb to a utilitarian scheme that is ungrounded, if not incoherent. Thus, for example, it would be wrong to go to war in the first place if policy exists that violates *jus in bello* from the outset.[37] We are justified, then, in arguing that the unity and integrity of the just war ethic are realized – and concretized – through the criterion of right intention. Right intention unifies *ad bellum* and *in bello* considerations. It reminds us that ends and means must agree. Morally calibrated means have the effect of "verifying" the justness of an interventionary cause.

In the classical just war tradition, right intention can be traced from Augustinian thinking through Aquinas to early-modern just war theorists such as Vitoria, Suárez, and Grotius, resurfacing again, in modified ways, in post–World War II thinking. Augustinian reflection on war is anchored in a full-orbed understanding of charity. Charity is not a mere emotion; rather, it expresses human volition and intention. All people possess and mirror this faculty. Charity is central to Augustinian thinking about coercive force, for charity is the *summum bonum*;[38] as it must inform *all* that we do, even going to war. For this reason, in his letter to Boniface, Augustine writes: "Be, then, a peacemaker even while you make war, that by your victory you may lead those whom you defeat to know the desirability of peace."[39]

Aquinas clearly enunciates the character of intention: "The way a moral act is to be classified depends on what is intended, not what goes beyond such an intention, since this is merely incidental."[40] In evidence thereof, it is significant that Aquinas's treatment of war in the *Summa* is contextualized, not under the rubric of justice but of *caritas.* This in no way minimizes Aquinas's concern for justice as a cardinal virtue; it only magnifies what lies at the heart of classical just war thinking: charity is active in nature – a 'principle of action' – and agrees with the basic conditions of justice. Similarly, Suárez addresses the subject of war as a duty of charity. This, along with his belief that the laws of war are binding on all nations, forms the main argument of his treatise on war found in *De Triplici Virtute Theo-logica, Fide, Spe, and Charitate* (*On the Three Theological Virtues, Faith, Hope, and Charity*). Charity is best observed in human action, and by its active nature, it might rebuke, punish, even permit going to war (though with strict demands). In the conduct of war, neighbor-love prevents us from pressing upon the adversary unrestrained violence. While severity is compatible with justice and charity, cruelty and barbarism are not. Authentic justice entails retribution, but it is guided by human dignity and charity.

As a moral criterion, right intention concerns the inner quality of an agent's action. Human beings can intend either good or evil. While we may grant that measuring the 'inner quality' of various acts can be difficult, it is nevertheless not impossible. In fact, at least in relatively democratic contexts, public acts must be reasoned, then explained and *justified*; we call this sort of procedure 'policy,' do we not?[41] Wrongdoing can be determined externally and juridically, just as guilt and innocence allow us to apply the criteria of discrimination and proportionality in the conduct of war.[42] Several characteristics define the essence of right intention: (1) it limits the agent to pursuit of the avowed cause, (2) it requires the agent always to have in mind the ultimate objective of a just and lasting peace, and (3) it necessitates charity, which views people as fellow human beings.[43]

Intention is based on voluntary action and reasons for acting. It aims at an outcome and thus proceeds through moral evaluation. Natural law moral reasoning causes us to conform to moral truth, as best we understand it, and not mere legal casuistry.[44] The principle of 'double effect' well illustrates the matter of intention as it applies to warfare and *in bello* considerations. There exists a fundamental moral difference between intending to harm or kill and observing side-effects of an action that does not intend to harm or kill. There is an objective 'structure' that can verify proper intention and thus inform proportionality and discrimination.[45] The principle of double effect, then, constrains human intention, assisting us in determining if certain moral conditions are met.[46] Apart from the natural law, the principle of double effect would be unintelligible and meaningless, since voluntary actions are guided by moral norms.[47]

What is so important – and distinct – about right intention is that it asks *why* we act. Intention unites ends and means. An agent can have wrong intentions even when a cause is deemed just. This, of course, is Aquinas's burden: "For it may happen that the war is declared by the legitimate authority, and for a just cause, and yet be rendered illicit through a vile intention."[48] And what intentions would be considered vile? Responses such as a cruel thirst for vengeance, a passion for inflicting harm, the lust to dominate, and an implacable hatred.[49] This is one important reason why consequentialism fails; it is morally bankrupt and incompatible with natural law moral reasoning. Helping to order a proper intention, in accordance with just war moral reasoning, is the aim to furnish a justly ordered peace. We aim to protect and preserve, where possible, concord and human community. Peace is diminished by any attitude, any intention, that ignores human dignity and the common social good. In the just war tradition, neighbor-love, a "Golden Rule" ethic, and *caritas* are the motivation and source of a truly just peace.[50]

Aquinas's discussion of the three *ad bellum* criteria – legitimate authority, just cause, and right intention – in Question 40 of the *Summa* (II-II) – suggests, if not demands, the importance of right intention as a qualifying and justifying agent. That discussion proceeds under the assumption that, *taken separately*, the three criteria are not sufficient causes to go to war. Only *together* can justification be reasoned. In addition, Aquinas's formulation, which has been standard in discussions of just war up to the present, would likely have been dependent on earlier (or roughly contemporary) discussions. For example, Alexander of Hales (d. 1245) had identified six criteria that distinguish a just from an unjust war: *auctoritas* (authority), *affectum* (state of mind), *intentio* (intention), *conditio* (condition), *meritum* (merit), and *causa* (cause).[51] Aquinas's understanding of right intention appears to accord with Alexander's *justum affectum* and *debita intentio*.[52]

One of the enduring contributions of Grotius' treatment of war in *De Jure Belli ac Pacis* is his extensive development of *ad bellum* criteria to *in bello* requirements. Writing at a time when war was widespread in Europe (the Thirty Years' War), Grotius saw the need to limit and restrain war.[53] He states what for him is a moral principle: "In the moral field, the means that lead to an end derive their inherent character from that end."[54] For this reason, he painstakingly attempted to specify what means of prosecuting war are just and unjust and to catalog and qualify just and unjust causes of war. Not merely content to expose unjust causes, Grotius sought to identify and expose dubious or wrong rationalizations, many of which suggested to him the need for mediation and arbitration. The structure of *De Jure* makes quite clear to the reader that not only just cause and legitimate authority but right intention informs the core of just war thinking. This is the age-old issue; the matters of *how* and *why* lie at the heart of just war moral reasoning.

A final thought on right intention is in order here, which underscores its centrality to just war thinking. Classic just war thinking refuses to divorce justice and charity; the two are not at odds, as they typically are in modern and contemporary thought.[55] To divorce these two virtues is to do irreparable damage to the character of both as well as to alter the very moral foundation upon which just war thinking rests. Justice will always seek to render "what is due" (the classical definition) to human beings and properly order human community,[56] while charity will always seek the means by which to respect human dignity and uphold justice.[57] Justice and charity meld in the just war tradition, and this union can be observed at three distinct levels. At one level, it is both just and charitable to *the criminal element* to prevent them from doing evil. At another level, that hindering is both just and charitable to *those in society or the community of nations* who are watching. Finally, that hindering is just and charitable to *potential offenders* who might be tempted in the future to do evil and undermine the common good.

Relatedly, justice and charity together distinguish between the criminal and the retributive act. Retribution, understood properly, has upper and lower limits. Negatively, it acknowledges the moral repugnance of draconian punishment for petty crimes as well as light punishment for heinous crimes. Positively, it has as its goal a greater social good and takes no pleasure in employing coercive force to guard the common good. Retributive justice thus serves a civilized culture, whether in the domestic or international context.[58] In the end, charity and justice are not at odds; rather, they are two sides of the same ethical coin.[59]

For war to be an instrument of wise policy, right intention – which guides *in bello* considerations and is undergirded by justice and charity (properly understood) – is integral.[60] Its importance can be summarized succinctly, if not neatly: if war is fought with the proper intention, over time even adversaries can become allies, as modern history – witness both Germany and Japan – indicates.[61] Again, it needs emphasizing at the most fundamental level that neither natural law thinking nor just war moral reasoning knows any division or separation between *ad bellum*, *in bello*, and *post bellum* justice. All three realms are organically unified. No considerations, whether in justifying coercive intervention, in conducting that intervention, or in the aftermath of such intervention, divorce themselves from the moral realities that govern life in this universe – among human beings, among communities, or among the community of nations.

Concluding Reflections

In sum, as a metaphysical notion, natural law mirrors the very order of things in all of creation, including human nature and human purposes. It serves to counter not only the biological reductionism, legal positivism, and utilitarian thinking of our era but also the extreme individualism that characterizes Western culture in general. It is the time-tested basis of human moral reflection, it reminds all people everywhere, regardless of social location, of moral reality that transcends human experience and social-cultural differences, and in the end, it points toward a Creator – a just Creator, who by nature forbids injustice.[62]

In natural law moral reasoning, law and morality are not divorced, as modern and postmodern thinking is inclined to do. Because of its baseline commitment to pursue good and avoid doing evil and its concern for the neighbor (as embodied in a "Golden Rule" ethic), natural law moral reasoning alone safeguards basic human freedoms and rights, thereby permitting human flourishing. It is thus a guide for social and military policy, and hence, for responsible statecraft. For this reason, then, at the end of Book III of *De Jure*, Grotius offers this prayer for leaders:

> May God, who alone can do it, inscribe these principles on the hearts of those in whose hands lies the government of the Christian world, and implant in them a mind which understands divine and human law, and which always remembers that it was chosen as a servant to rule man, the creature best loved by God.[63]

In the end, natural law thinking underpins and informs the entire just war tradition in its various component parts. The two traditions are intimately bound up with one another, given the shared assumption that there are universally binding moral obligations that transcend culture and social location – obligations that are pre-political and pre-legal. The natural law furnishes the basis for thinking about humanitarian and military necessity – which are not polar opposites – insofar as it governs military intentions and hence how war is to be conducted. War is just – based on natural law moral intuitions and just war moral reasoning – where and when it seeks to protect and defend the relatively innocent.[64] Humanitarian and military intervention – the two always go together – proceed from the common dignity of all humans. Therefore, we rescue those in oppressive trauma and need; such is both 'natural' and 'law.' As ancient wisdom reminds us: "Rescue those being led away to death; hold back those staggering toward slaughter. If you say, 'But we knew nothing about this,' does not he who weighs the heart perceive it? Does not he who guards your life know it? Will he not repay everyone according to what they have done?" (Prov. 24:11–12 NIV). Both moral traditions, then, unify any and all considerations informing whether or not to go to war (*jus ad bellum*) with how to conduct war (*jus in bello*) and how to approach war's aftermath (*jus post bellum*).

In short, the same understanding of human nature that permits coercive intervention in human affairs places limits as to how that intervention is to be done. That understanding of permission and restraint provides justification for conflict when and where belligerents have as their aim the protection and defense of fellow human beings and the restoration of a justly ordered peace.

Notes

1 Thomas Aquinas, *Summa Theologica* I-II Q. 94. John Courtney Murray has stated it well: "The question of [war's] proportion must be evaluated … from the viewpoint of the hierarchy of *strictly moral values*… There are greater evils than the physical death and destruction wrought in war. And there are human goods of so high an order that immense sacrifices may have to be borne in their defense" (*We Hold These Truths* [New York: Sheed and Ward, 1960], 261, emphasis added). Moral discernment, of course, becomes exceedingly difficult in a society that struggles to identify – let alone distinguish between – good and evil.
2 James Turner Johnson is one of the few to establish this linkage; see "Natural Law as a Language for the Ethics of War," chapter 4 of *Just War Tradition and the Restraint of War* (Princeton, NJ: Princeton University Press, 1981), 85–118.
3 Jean Bethke Elshtain, "The Just War Tradition and Natural Law," *Fordham International Law Journal* 28, no. 3 (2004): 742–755, cites Augustinian thinking to help anchor our understanding of human "nature": it combines our fallenness and our dignity, and hence, our *solidarity*.
4 Thomas Aquinas describes the natural law as "the rational creature's participation" in "the eternal law" of God (*S.T.* I-II Q. 91, a. 2), whereby good must be done and evil avoided. Classical thought, with Aquinas, holds the natural law to be unchanging, universal, and independent of cultural particularities. In his encyclical *Splendor of Truth*, John Paul II writes: "Whereas the natural law discloses the objective and universal demands of the moral good, conscience is the application of the law to a particular case; this application of the law thus becomes an inner dictate for the individual, a summons to do what is good in this particular situation" (no. 77). This, of course, counters the contemporary view of criminal behavior as mental or biological deficiency, and hence with no attendant moral culpability, whereby the meaning of virtue, character, and the common social good disappear. It also counters (to some extent) utilitarian ethics, which is often the chief mode of decision-making in the military. I grant here that I am generalizing.
5 *De Jure Belli ac Pacis* (*The Law of War and Peace*), Proleg. 11. Grotius accompanies this assertion, however, with a qualifying comment: only "the greatest wickedness,"

he notes, can claim that God does not exist or that he is uninvolved in human affairs. Rather, Grotius insists that we owe the Creator our obedience in all things.

6 A graphic example of deficient theology comes to us in the volume *Faith and Force: A Christian Debate about War* (Washington, DC: Georgetown University Press, 2007). Co-author David L. Clough writes: "I don't think that we need to believe that there's a natural law to realize that we need international agreement ... my main concern with natural law thinking is that it starts us off by looking for moral norms that will be universally applicable, whereas I want to start by listening for what the Church that I belong to is called to do" (31–32). Did you catch that? 'My church,' *not the combined wisdom of the Christian moral tradition*, should guide us. This from a professor of theological ethics at the University of Chester (in the UK), and the author of *Ethics in Crisis* (Routledge, 2017). Clearly Christian ethics *would* seem to be in an advanced state of crisis.

7 This is true even when early-modern just war theorists were more explicit in applying natural law thinking to war in cross-cultural contexts (on which, see below).

8 This corresponds with the *tranquillitas ordinis* described by Augustine in *City of God* 19.12–13. In the present age, this *saeculum*, peace must be guarded by the political act, whereby law must regulate the behavior of human beings and parties. Thus, for example, as a result of criminal war-making, judgment by governing authorities must be the necessary response. That is, if the advancing of war is a crime (such as in Russia's war with Ukraine, in our day), armed conflict is justifiable and needed in response.

9 This is precisely Augustine's argument, as developed in *City of God* 19.12: "Take even a band of highwaymen. The more violence and impunity they want in disturbing the peace of other men, the more they demand peace among themselves. Take even the case of a robber so powerful that he dispenses with partnership, plans alone, and singlehandedly robs and kills his victims. Even he maintains some kind of peace, however shadowy, with those he cannot kill and whom he wants to keep in the dark with respect to his crimes." Here I am dependent on the translation found in Saint Augustine, *The City of God*, ed. Vernon J. Bourke and trans. Gerald G. Walsh et al. (Garden City: Image Books, 1958), 452.

10 For this reason, in the opening chapter of *Mere Christianity*, titled "Right and Wrong as a Clue to the Meaning of the Universe," C.S. Lewis poses the rhetorical question, "Why is it that all people – everyone, everywhere – complain when injustice visits them?" (my rough translation). Indeed, Lewis concludes, all people intuit the reality of justice and injustice, even when they are in denial of its theoretical and moral foundations.

11 If we jettison a proper view of human nature, "all bets are off" in terms of morality and justice, as Elshtain warns (See Note 1.). Without a proper view of human nature, justice becomes fluid, with immediate and catastrophic implications for social and public policy, both domestically and internationally.

12 J. Budziszewski, *What We Can't Not Know: A Guide* (Dallas: Spence, 2003).

13 Rom. 2:14–15.

14 The natural law tradition has a long, rich history, extending from Heraclitus in Greek philosophy through the early church fathers to Thomas Aquinas to early-modern thinkers (including the magisterial Protestant reformers). Obscured somewhat by Enlightenment thinking, natural assumptions resurface after World War II, undergirding international conventions on human rights.

15 Whether or not people *live up* to that norm is quite a different matter.

16 This third expression of the "Golden Rule" demonstrates the folly of ideological pacifism, which, as G.E.M. Anscombe famously noted, makes the world unsafe for all. Clearly, as the natural law tradition, the just war tradition, and the Christian moral tradition together confirm through the ages, we are *not* to "turn the other cheek" of an innocent third party when an aggressor threatens. Shared dignity and shared morality require that we defend and protect that third party.

17 Calvin writes that "there is nothing more common than for a man to be sufficiently instructed in a right standard of conduct by natural law" (John Calvin, *Institutes of the Christian Religion*, ed., John T. McNeill, trans. Ford Lewis Battles [Philadelphia: Westminster, 1960], 2.2.22). In his writings, Luther, for whom grace was so crucial, assumes that the natural law and an uneasy conscience are the first point of contact between human beings and their Creator.

18 I shall never forget an incident that occurred at an ethics symposium at the US Army's Command and General Staff College at Fort Leavenworth, Kansas, some years ago. I was presenting the close relationship between natural law thinking and just war thinking, and observed the importance of acknowledging *the* foundation of *transcendent* moral reality as illustrated through the "Golden Rule." In the question-and-answer session, which was quite lively, a Major stood up, thanked me for my talk, and then proceeded to vent his frustration by complaining, "Sir, the Army tells us constantly that we should *be good* and *do good*, but it doesn't tell us *why!*" How sad; "it doesn't tell us *why*." Can we ask that our soldiers be willing to sacrifice their lives without telling them *why*? Relativists, it goes without saying, cannot be good soldiers. It is absurd to assume that people can and will sacrifice themselves to ethical relativism.

19 James Turner Johnson, "Natural Law as a Language for the Ethics of War," *Journal of Religious Ethics* 3, no. 2 (Fall 1975): 217–242, repro. in idem, *Just War Tradition and the Restraint of War* (Princeton, NJ: Princeton University Press, 1981), 85–118. Johnson identifies three basic ways in which people typically understand 'natural law': (1) as a past, medieval notion; (2) as an unnecessarily value-laden idea; and (3) as a reality that is hidden yet accessible. I would add a fourth: as a "Catholic" thing. Johnson prefers to understand the natural law as hidden though accessible to people.

20 For Christian believers, an important question confronts us. Shall we use inherently religious language in the public sphere, on the one hand, or should we abdicate Christian values in the public sphere in order to have a place? Our answer must be neither. Natural law thinking allows us to bridge the 'sacred' and the 'secular' by appealing to what transcends particular social, cultural, and political differences or limitations. It assumes commonality, and it assumes universal accessibility through reason based on a shared human 'nature.'

21 Vitoria understands this "law to the nations" to be established by "natural reason" (*De Indis* [*On the American Indians*] Question 3, Art. 1). What escapes the modern reader is the fact that Vitoria was stunningly adamant in rejecting Spanish imperial pretensions of his day. What strikes the reader is Vitoria's attention to *why* we go to war, not just *whether*. Vitoria insists that the aim of war must be "peace and security" and "the good of the whole world" (*De Jure Belli* 1.4 and 3.9). Suarez's overriding concern in his treatment of war was how states conduct themselves; hence, his treatment of war under the rubric of charity and the natural law, which he insisted are common to all people everywhere. Just cause, right intention, and legitimate authority constitute the core of Grotius' understanding of just war. Grotius writes that human nature is "the mother" of natural law and thus the "grandmother" of civil law (*De Jure Bello ac Pacis,* Proleg. 16).

22 A common objection, particularly among Protestant thinkers, is that natural law thinking is autonomous to a so-called Christian ethics and to the authority of Scripture. This view is utterly false and mirrors a fundamental misunderstanding both of law as an entity and of natural law specifically. The Hebrew and Christian scriptures presuppose the natural law. Moreover, the "Ten Commandments" simply define the contours of the natural law as a moral code that applies to all of humanity – a code that is not exclusive to any particular culture or society.

23 Francisco Suárez, *Three Theological Virtues* 3.8.1. I am here dependent on the English translation provided in James B. Scott, *The Spanish Origin of International Law: Lectures on Francisco de Vitoria (1480–1546) and Francisco Suárez (1548–1617)* (Washington, DC: Georgetown University Press, 1929), 77.

24 Suárez, 3.8.2.

25 Grotius, *De Jure Belli ac Pacis*, Proleg. 30.

26 Much contemporary literature on 'just war' is written by individuals who, philosophically, stand *outside* of the just war tradition. Two tell-tale signs help identify such writers/theorists: (1) a wholesale neglect or ignoring of the natural law tradition as it informs just war moral reasoning and (2) the tendency to speak in terms of 'just war theory' rather than the just war *tradition*.

27 Hence, the logic of both natural law and just war moral reasoning (classically understood) prevents one from holding the position that both sides of a war can be just. And if both sides were 'just,' then it would be immoral and 'unlawful' to fight, whether in offensive or defensive terms, and 'just cause' would be rendered meaningless. It is for this reason, then, that Grotius favored arbitration. Only one party, he insisted, could be justified in war (*De Jure Belli ac Pacis* II.23.13); both sides cannot conduct "just retribution" and serve the goal of an enduring and just peace if both are morally justified and externally legitimate.

28 Anthony Coates, "Is the Independent Application of Jus in Bello the Way To Limit War?," in David Rodin and Henry Shue, eds., *Just and Unjust Warriors* (Oxford: Oxford University Press, 2008), 181.

29 *De Jure Belli ac Pacis* III.1.2

30 'Military necessity,' of course, might mean "'death to the infidel' – not only in a medieval context but in the twenty-first century where radicalized Islamist terrorism justifies all means necessary" – or it might mean utilizing military means that force the adversary to surrender without killing innocent people and destroying the environment needlessly.

31 Michael Walzer, *Just and Unjust Wars* (New York: Basic Books, 1977), 251–268. Among those objecting to Walzer's reasoning, James Turner Johnson, *Just War Tradition and the Restraint of War* (Princeton, NJ: Princeton University Press, 1981), 25, responds by arguing that Walzer's mistake is to absolutize the enemy threat.

32 Walzer, *Just and Unjust Wars*, 268.

33 Justice by definition indicates limits; otherwise, there is *no such thing* as 'justice' – in *any* context. At the most fundamental level, human law cannot permit what the natural law forbids, especially in war (so Grotius, *De Jure Belli ac Pacis* II.2.5 and III, 4.2, 15).

34 It is part of their nature that mass murderers, despots, dictators, and revolutionaries – we need only remember Stalin, Hitler, Mao, Pol Pot, Che Guevara, Idi Amin, and others in more recent history – eschew restraint. For them, anything and everyone can be sacrificed for the cause.

35 Indeed, military necessity justifies a resort to all measures that are necessary for securing submission of the adversary, provided that these measures are not inconsistent with moral criteria of discrimination and proportionality. Moral obligations in military service entail both to act and to refrain from acting. We may say that proportionality includes and extends the notion of military necessity.

36 War crimes and crimes against humanity violate the natural law, at the most basic level, and the just war criteria of right intention, military necessity, proportionality, and discrimination.

37 So C. A. J. Coady, "The Status of Combatants," in David Rodin and Henry Shue, eds., *Just and Unjust Warriors* (Oxford: Oxford University Press, 2008), 158.

38 Augustine's discussions of the priority of charity are scattered – found in *Confessions*, *City of God*, and various letters. Perhaps familiar to most people are his famous words "Love, and do what you will" (*Homilies on the First Letter of John* 7.8).

39 "To Count Boniface" (#189), in Saint Augustine, *Letters – Volume IV* (New York: Fathers of the Church, 1955), 269. Aquinas, in his treatment of right intention, will repeat the admonition of Augustine given to Boniface (*S.T.* II-II Q. 40, a. 1).

40 *S.T.* II-II Q. 64, a. 7; I am here reliant on St. Thomas Aquinas, *Summa Theologiae*, vol. 38 (New York and London: McGraw-Hill and Eyre & Spottiswoode, 1964), 41–42.

41 Joseph Boyle, "The Necessity of 'Right Intention' for Justifiably Waging War," in Anthony F. Lang, Jr., Cian O'Driscoll, and John Williams, eds., *Just War: Authority, Tradition, and Practice* (Washington, DC: Georgetown University Press, 2013), 185, defines right intention as "effective motivations" and not merely inclinations.

42 We apply the same moral reasoning to "criminal justice" in the domestic context. Guilt and innocence are determined by *public* acts, not internal thoughts or one's psychology.

43 So William V. O'Brien, *The Conduct of a Just and Limited War* (New York: Praeger, 1981), 33–35. O'Brien, accurately in my view, calls this criterion "elusive but important" (338).

44 We call – or at least we used to call – this movement in the direction of moral reality 'virtue' and 'character formation.'

45 Contra Brian Orend, *The Morality of War* (Peterborough, ON: Broadview, 2006), 46, who argues that just cause is "objective" and right intention is "subjective."

46 Walzer, *Just and Unjust Wars*, 155, argues that the 'principle of double effect' is valid if based on a '*double* intention': first, that the good be achieved, and second, that foreseeable evil be reduced as far as possible.

47 Few have examined the matter of motive and intention more clearly than Joseph Boyle, in "Just War Thinking in Catholic Natural Law," in John A. Coleman, ed., *Christian Political Ethics* (Princeton, NJ and Oxford: Princeton University Press, 2008), 217–231, esp. 222–227.

48 *S.T.* II-II, Q. 40, a. 1, repro. in Reichberg, et al., eds., *The Ethics of War*, 177. Anthony J. Coates, *The Ethics of War* (Manchester: Manchester University Press, 1997), 161, rightly observes that "the moral efficacy of just cause is dependent upon the existence of right intention."

49 These vices are noted by both Augustine and Aquinas. Here Aquinas, in *S.T.* II-II Q. 40, is citing Augustine's *Contra Faustum* 22.74.

50 Instructive is Aquinas's treatment in the S*umma* of those vices that are opposed to *caritas* (*S.T.* II-II QQ. 34–43). Similarly, see Suarez's discussion in *De Triplici Virtute Theologica*, Disputation XIII ("*De Caritate*").

51 See "Alexander of Hales (ca. 1185–1245): Virtuous Dispositions in Warfare," in Reichberg et al., eds., *The Ethics of War*, 156–158.

52 So Gregory M. Reichberg, "The Decision To Use Military Force in Classical Just War Thinking," in Johnson and Patterson, eds., *Ashgate Research Companion to Military Ethics* (Surrey, and Burlington: Ashgate, 2015), 16.

53 According to Richard Shelly Hartigan, *The Forgotten Victim: A History of the Civilian* (Piscataway: Transaction, 1982), 90, war during the sixteenth century claimed roughly 1.6 million lives.

54 *De Jure Belli ac Pacis* III.1.2. I am here relying on the translation found in *The Law of War and Peace*, p. 269 (See Note 3.).

55 This tragic divorce characterizes both secular and religious approaches to ethics. Consider, for example, a representative case of the latter: Timothy P. Jackson, *The Priority of Love: Christian Charity and Social Justice* (Princeton, NJ and Oxford: Princeton University Press, 2003).

56 The just war tradition, it needs reiterating, gives an account of *comparative* rather than absolute or perfect justice, as Jean Bethke Elshtain, *Just War Against Terror: The Burden of American Power in a Violent World* (New York: Basic Books, 2003), 53, reminds us.

57 Benedict XVI, in his 2009 encyclical *Caritas in Veritate*, expresses well the resultant calamity when love is divorced from truth and justice: "Without truth, charity degenerates into sentimentality. Love becomes an empty shell, to be filled an arbitrary way. In a culture without truth, this is the fatal risk facing love. It falls prey to contingent subjective emotions and opinions, [and] the work 'love' is abused and distorted, to the point where it comes to mean the opposite" (no. 3).

58 Consider the comments of the former prosecutor of the International Criminal Tribunals for the former Yugoslavia and Rwanda, in a speech delivered at the U.S. Holocaust Museum: "Where there have been egregious human rights violations that have been unaccounted for, where there has been no justice, where the victims have not received any acknowledgement, where they've been forgotten, where there's been a national amnesia, the effect is a cancer in society." See Richard Goldstone, "Healing Wounded People," *The Washington Post* (February 2, 1997), C4. Vitoria would agree: just retribution both restores and deters; without fear and deterrence, the adversary will continue with even worse injustice and oppression (*De Jure Belli* 1).

59 Elsewhere I have developed the argument for the unity of justice and love more extensively. See, for example, J. Daryl Charles, "Justice, Neighbor-Love and the Just-War Tradition: Christian Reflections on Just Use of Force," *Cultural Encounters* 1, no. 1 (Winter 2004): 47–67, and idem, "The Natural-Law Underpinnings of Religious Freedom – A Closer Look: Justice and Neighbor-Love in Symbiosis," in *Natural Law and Religious Freedom: The Role of Moral First Things in Grounding and Protecting the First Freedom* (Oxon, and New York: Routledge, 2018), 177–228 (Chapter 6).

60 Modern and contemporary discussions tend to neglect right intention. Even Michael Walzer's *Just and Unjust Wars* pays scant attention to it. Remarkably, the *Catechism of the Catholic Church* totally ignores it in its treatment of war (see nos. 2307–2317).

61 While some might object to this statement, O'Brien states the matter well: the United States and the Allies "overcame their delinquencies" (e.g., bombing German cities) as *post bellum* justice in time was realized. Just war theorist Darrell Cole, in responding to the question of how to determine if right intention is satisfied, answers that it is identified by observing a belligerent's acts both during and after a conflict; see Darrell Cole, "War and Intention," *Journal of Military Ethics* 10, no. 3 (2011): 174–191.

62 Several generations removed, Dorothy Sayers, in her classic work, *The Mind of the Maker* (New York: Harcourt Brace & Co., 1941), spoke forthrightly: "Defy the commandments of the natural law, and the race will perish in a few generations; co-operate with them, and the race will flourish for ages to come. That is the fact; whether we like it or not, the universe is made that way. This commandment is interesting because it specifically puts forward the moral law as the basis of the moral code: because God has made the world like this and will not alter it, therefore you must not worship your own fantasies, but pay allegiance to the truth" (12).

63 Grotius, *The Law of War and Peace* II.25.8.

64 For this reason, both moral traditions distinguish between the state and a society, since the belligerent typically is the state. For example, post–World War II deliberations and activity were aimed at eliminating Nazism, not German society.

3 Military Necessity as a Distinct *Jus in Bello* Principle

A Classical Just War Perspective[1]

Christian Nikolaus Braun

Introduction

The principle of military necessity is one of the principles of the *jus in bello* code of the Law of Armed Conflict (LOAC), besides those of distinction, proportionality, and humanity.[2] During my time as a senior lecturer at the Royal Military Academy Sandhurst, my colleagues and I taught Officer Cadets the following reminder: Military Necessity, Distinction, Proportionality and Humanity = Moral Decisions Prevent The Hague (or MDPH = MDPH). While this reminder was effective in helping Cadets remember the four LOAC principles, it was deficient in one respect. After all, *moral* decisions are not necessarily *legal* decisions. In the context of teaching LOAC, the legal dimension was of course the main concern, although LOAC cannot sufficiently be understood without a grasp of its genesis, which is inseparably linked to the moral argument. More importantly still, the soldier who receives training in the morality of war will find in it a support system that helps him or her apply LOAC in the right way. After all, soldiers may encounter circumstances where morality may require them to take on more risk to themselves than LOAC requires.

This interplay between the morality and legality of war turns the attention to just war thinking, a rich tradition that operates from what has been described as a dual theme of permission and restraint.[3] This dualism applies equally to the decision to go to war (*jus ad bellum*) and the decision to use force in war (*jus in bello*). Interestingly, contemporary just war thinking commonly deviates from the list of LOAC principles. Most current accounts concentrate on the *jus in bello* principles of discrimination and proportionality only. Consequently, the principle of military necessity receives relatively little treatment by the majority of contemporary just war thinkers.[4] There is thus reason to ask why military necessity is a fundamental principle in LOAC, whereas just war thinkers apparently do not see it in this way. Even more importantly, would just war thinking benefit from a distinct *jus in bello* principle of military necessity or do the two established principles sufficiently regulate just conduct?

In this chapter, I will answer these questions by turning to the classical just war as found in the thought of Saint Thomas Aquinas (1224/5–1274). Nowadays, one often finds a division of just war into so-called Walzerians and their revisionist

DOI: 10.4324/9781003390398-3

critics.[5] Considerable ink has been spilled on the intradisciplinary disagreements of a field that has become "confusingly polarized"[6] and that seems to be engaged in what I call "the fight for the just war tradition."[7] In this metaphorical fight, both sides claim to advocate the accurate interpretation of just war and the exchange between the two camps on substantive issues has been very limited only. However, whilst Walzerians and revisionists disagree on several key questions in the ethics of war, they seem to be united in what Eric Patterson calls the general avoidance of military necessity.[8] Crucially, the classical just war of Aquinas went even further than the two infighting contemporary camps. Whilst Walzerians and revisionists list the *jus in bello* principles of discrimination and proportionality, Aquinas did not provide a standalone account of just conduct in war at all. However, the absence of distinct principles of discrimination and proportionality – and military necessity – must not be taken to mean that Aquinas was not interested in just war conduct. By turning to Aquinas, I will be able to highlight a crucial concern regarding military conduct that both contemporary camps miss, namely, the significance of virtue. Both Walzerians and revisionists emphasize the importance of rules for just war conduct. Aquinas, in contrast, had a different starting point. Whilst St. Thomas did not discount the importance of rules, he was primarily interested in virtuous behavior. That is the reason why he did not develop an elaborate *jus in bello* code. As I will argue in this chapter, the Thomistic contribution to the debate about military necessity and its place in just war thinking is to highlight the importance to turn soldiers into virtuous actors. Aquinas accepted that what we today call the *jus in bello* principles can justifiably be violated in war. This acceptance amounts to an affirmation of military necessity. Crucially, however, such violations must never happen intentionally. Grounded in Aquinas's understanding of virtue, I will argue that rules can help soldiers in making the right decisions on the battlefield, but they also need to be virtuous to apply these rules correctly. Put differently, the question of whether we may want to add a distinct *jus in bello* principle of military necessity is to some extent a secondary concern. Training in the virtues comes first.

This chapter's argument unfolds in three steps. Firstly, I will engage with the military necessity principle as found in the classical *bellum justum*. I will point out that military necessity in the above understanding was an important consideration in the Middle Ages. In particular, I will discuss the divergent attempts to regulate war conduct advocated in canon law and the code of chivalry. Next, I will turn to Aquinas's understanding of virtue and explain why he gave great emphasis to virtuous behavior in war and did not put forward a detailed *jus in bello* code. Building on the crucial place for virtue in Aquinas's just war I turn, in the conclusion, to the debate about the right place of military necessity in today's just war *jus in bello*. I will argue that the main contribution the Thomistic just war can make in this regard is to create virtuous soldiers that will naturally be able to make the right decisions in the heat of battle and, therefore, apply military necessity correctly.

Military Necessity and the Classical *Bellum Justum*

From the point of view of the classical just war, just and unjust combatants would not be each other's moral equals.[9] Even if they fought according to the established rules of war, unjust combatants would still be taking part in acts of wrongdoing, which distinguished them from their counterparts on the just side. As a result, the classical just war was all about delimiting when and under what restraints the just side would be justified to employ force. This clear-cut distinction between the just and unjust side, according to Stephen Neff, had profound consequences for the consideration of military necessity:

> For the overwhelming part, just-war doctrine, in its treatment of the conduct of hostilities by the just side, remained resolutely lodged at the level of broad general principle, and indeed of only one general principle at that: necessity. That meant that the just side was permitted to use whatever degree of force was strictly necessary in the particular circumstances of the case to bring about victory. Beyond that point, *all* force became unlawful.[10]

Neff calls the natural-law based understanding of the necessity principle "Janus-faced," because at times it could be employed for moderation's sake, whereas at other times it could be used to justify extraordinary force to secure victory.[11] Suggesting that the classical just war understanding of necessity included a "balancing test" that weighed the suffering caused by a war against the military advantage brought about by a particular operation he cites the following passage from Francisco de Vitoria's (1492–1546) *On the Law of War*:

> Care must be taken [cautioned Vitoria] to ensure that the evil effects of the war do not outweigh the possible benefits sought by waging it. If the storming of a fortress or town garrisoned by the enemy but full of innocent inhabitants is not of great importance for eventual victory in the war, it does not seem to me permissible to kill a large number of innocent people by indiscriminate bombardment in order to defeat a small number of enemy combatants.[12]

At first look, Neff's take seems to be an accurate reflection of the classical just war. Clearly, the passage from Vitoria he cites affirms that the shedding of innocent life can be justifiable in war. However, the way Neff puts his argument is misleading, because he does not mention a fundamental aspect in Vitoria's argument and, in fact, all thinkers – classical and contemporary– who work in the Catholic tradition.[13] While the killing of the innocent in war may at times be morally justifiable, Vitoria emphasized that it must never be *intentional*: "First, it is never lawful in itself intentionally to kill innocent persons."[14] This prohibition of intentional killing of the innocent also underpins Vitoria's thought in the passage that comes immediately before the one Neff employs to make his point on military necessity in the classical just war:

> Second, it is occasionally lawful to kill the innocent not by mistake, but with full knowledge of what one is doing, if this is an accidental effect: for example, during the justified storming of a fortress or city, where one knows there are many innocent people, but where it is impossible to fire artillery and other projectiles or set fire to buildings without crushing or burning the innocent along

with the combatants. This is proven, since it would otherwise be impossible to wage war against the guilty, thereby preventing the just side from fighting.[15]

Undergirding this reasoning is the Doctrine of Double Effect (DDE) that is traced to Aquinas and that Vitoria, as a Thomist and fellow Dominican, was naturally applying in his thinking on war.[16] This doctrine consists of two parts. Firstly, it holds that unjust deeds may not be carried out intentionally, regardless of the positive results the wrongful action may bring. Secondly, as the flipside of the first part, DDE accepts that just deeds may come with foreseeable negative side effects. Additionally, the agent carrying out the good action is not morally required to abstain from his/her doing to prevent the negative side effect.[17] Only as part of this understanding of ruling out the intentional killing of innocent persons can considerations of what we today call military necessity be found in the classical just war. It needs to be remembered that a just war essentially constituted an act of law enforcement that unfolded within a legal frame. As Gregory Reichberg puts it succinctly:

> The judicial focus entailed that only those discernibly responsible for wrongdoing could be directly targeted with the harms of war. By drawing a distinction between offenders (*nocentes*) and the innocent (*innocentes*), the requirements of discrimination and proportionality were made applicable to the treatment of enemy subjects in a just war.[18]

Crucially, arguing that the intentional killing of innocent persons is irreconcilable with the classical *bellum justum* does not mean that this constraint was commonly observed. As students of history know, the brutality of medieval warfare functions as a testament to the fact that oftentimes all belligerents thought to be fighting a just war and also disregarded the prohibition on intentionally killing the innocent. It should also be mentioned that DDE is liable to abuse, as combatants can always wrongfully claim that the collateral damage they have caused was not intended. Of course, from a theological point of view, such action would be sinful, but also very difficult to prove. I will have more to say on this in the next section.

Medieval Church attempts to restrict war conduct, such as the Truce and Peace of God movements or the rules of the Second Lateral Council of 1139, marked an attempt to address the excesses of medieval warfare, but were of relatively little practical consequence.[19] For example, pointing to the poor observance of the Truce of God, which defined certain days as off-limits for fighting, James Turner Johnson identifies military necessity as one reason why Christian rulers tended not to follow it.[20] Generally, in medieval practice, the principles regulating war conduct emanated more from the chivalric code than the Church attempts of canon law or the thought of scholastics like Aquinas.[21] In the early canon law attempts to regulate war conduct, Johnson detects an idealism the code of chivalry did not exhibit. While the Church's argument emphasized the inviolability of what we today call noncombatant immunity, the knights operated from a more permissive reading that allowed them to live off the land:

> The rights of those deserving to be let alone because the war was not their war might exist as a moral absolute, yet though the phrase was not yet

coined, military necessity required that the knights take what they needed and attempt to deny what was left to their opponents.[22]

Put differently, Church regulations as they applied to war conduct could not sanction the intentional killing of innocent people, because it would violate a fundamental principle of the Christian religion. The fact that the code of chivalry did not embrace this understanding and, depending on the circumstances, justified the intentional and indiscriminate killing of innocent persons, would not change the sinfulness of such action in the eyes of the Church. Later, the two sources on noncombatant immunity, churchly canon law and the code of chivalry, would coalesce in a cultural consensus in the writings of Honoré Bonet (c. 1340–c. 1410) and Christine de Pizan (c. 1364–c. 1430). The Church was part of this consensus and now "its theology was also philosophy, and its canon law overlapped civil law."[23] Importantly, the fundamental principle of rejecting the intentional killing of innocent persons, which was in tension with the code of chivalry, made it into this consensus. Otherwise, the Church could not have embraced the consensus without sacrificing the prohibition on intentionally taking innocent life, which at the most fundamental level goes back to the Christian calling to do good and avoid evil. Not surprisingly, Aquinas started his definition of right intention from precisely this understanding: "Thirdly, it is necessary that the belligerents should have a rightful intention, so that they intend the advancement of good, or the avoidance of evil."[24] It goes without saying that the prohibition of taking innocent life had always been part of the Christian religion. The canon law attempt to designate certain groups as off limits was a specification of this general principle aimed at restraining the conduct of war. In this respect, it differed from the idea that undergirded the code of chivalry. Johnson distinguishes between noncombatant immunity granted by right and granted as a gift. The former applied to the canon law attempt and could not be sacrificed intentionally. The latter approach however, present in the chivalric code, was a voluntary constraint on war conduct that could be revoked if the knights found it *necessary* to do so.[25] In a sense, the chivalric understanding of military necessity has parallels with the nineteenth-century German doctrine of *Kriegsraison,* which, at the *jus in bello* level, held that any type of military conduct is justifiable if it is considered to be militarily necessary: "Kriegsraison geht vor Kriegsmanier;" when militarily necessary, the rules of war can be abandoned.[26]

Aquinas, *Jus in Bello* and Virtue

Although the necessity principle in the classical *bellum justum* was not as permissive as Neff suggests he is right in a related aspect, namely, the classical just war's emphasis of the *jus ad bellum* over the *jus in bello*. Aquinas as the arguably most influential medieval just war thinker did not spell out a sophisticated account of *jus in bello* in his treatment of just war in the *Summa Theologiae*.[27] In this respect, Aquinas was in line with medieval just war thinking generally, which had engaged with war conduct "only tentatively, and only in certain respects."[28] St. Thomas gave almost exclusive attention to his three *jus ad bellum* criteria of legitimate authority, just cause and right intention. The main reason for delegating the *jus in bello* to second place was the rationale that the foremost decision regarding war would be whether to employ force. The question of how to employ justified force came only after this initial decision.[29] That said, as a medieval churchman, Aquinas certainly approved of the churchly attempts to constrain warfare discussed above, all of which had happened before Aquinas was born. It should also be remembered that Aquinas's succinct systematization of the classical *bellum justum* relied heavily on canon law as found in Gratian's *Decretum* and the later work of the Decretists and the Decretalists. One could also argue, as Johnson does, that the two standard *jus in bello* criteria of contemporary just war, discrimination and proportionality, are inherent in Aquinas's *jus ad bellum* criterion of right intention. In sum, clearly, Aquinas was not against rules as such.

At the same time, however, in his discussion of right intention, Aquinas emphasizes the fundamental importance of cultivating the right moral inclinations.[30] Just as centuries later the Lieber Code would rule out acts of cruelty, Aquinas's just war has a built-in firewall against illicit motivations in war conduct.[31] Undergirding his understanding of *recta intentio* is an account of virtue. In fact, as Reichberg notes, Aquinas is the only scholastic who embeds his discussion of *bellum justum* within a typology of the virtues. The reason for this was the understanding that rules alone would not be sufficient to regulate war conduct. Due to the heat of battle, soldiers could be unable to take the time to reflect on the right application of the rules. That is why they depended on cultivating virtue, which would let them habitually do the right thing on the battlefield. In contrast, his successors Vitoria and Francisco Suárez (1548–1617) abandoned this virtue perspective, which is why they developed a *jus in bello* casuistry that was supposed to replace Aquinas's stress of the virtues.[32]

Reflecting on Aquinas's treatment of just war and virtue, Darrell Cole has made an original contribution that, crucially, can inform this volume's discussion about how to deal with military necessity in contemporary just war thinking.[33] Before I turn to Cole's interpretation of Aquinas, let me address the potential concern that Aquinas, as a Christian theologian, is a poor choice to argue about the military necessity principle today. Today's Western militaries are secular institutions and their members come from various religious backgrounds or have no such background at all. Cole identifies the virtues of justice and charity as most important in Aquinas's account of just war.[34] That does not mean that other virtues do not apply to war, but justice and charity are taken to be the most relevant. Crucially, the theological virtue of charity is the more important among the two. In Cole's words, it is "the crucial virtue in Aquinas's account of Christian soldiering."[35] Charity is the highest of the theological virtues because it leads the Christian to her ultimate telos, which is unity with God. It is this lodestar function that lets charity transcend the so-called natural virtues like the cardinal virtue of justice. Crucially, however, Cole argues that while Aquinas wrote his treatment of just war for Christians, that does not exclude the possibility of non-Christians fighting just wars. Put differently, the absence of Christian charity does not deny the existence of just wars: "Non-Christian leaders can make decisions about whether a proposed war is just without knowing the final end, for justice is a natural moral virtue, and just wars are fought for limited earthly goods (peace and order)."[36] The same would also apply at the *jus in bello* level. Both Christian and non-Christian soldiers could fight virtuously, with the former group's actions oriented toward a supernatural end and the latter oriented solely toward an earthly end.[37] As a result, Aquinas's account of virtue, also as a secular reading, can help us grapple with contemporary moral issues arising in war, including the question of how to deal with the principle of military necessity. In that sense, the ongoing (secular) revival of virtue ethics in the field of military ethics connects with Aquinas's account of *bellum justum*.[38]

Cole argues that contemporary accounts of just war tend to rely on rules as underpinning decision-making and these rules are derived from the established just war criteria. The problem he identifies in this procedure is that such "rule-driven" accounts are not sufficiently interested in how wars are actually fought.[39] As an example for such an "emphasis on rules with a correspondingly sad lack of virtue language," Cole points to the influential 1983 pastoral letter *The Challenge of Peace* that the US Catholic Bishops put forward.[40] Given the close relationship between Walzer's just war and international law, as well as revisionists' attempt to clearly define the correct moral behavior for oftentimes other-worldly scenarios, it seems fair to argue that Cole's critique applies to them, too. What is missing in accounts that do not rely on virtue, according to Cole, is that the overreliance on rules forecloses answers to the crucial questions of "how anyone can obey those rules and why they should want to do so in the first place."[41] Most revisionist thinking, for example, fails to recognize that in the heat of battle there is no time for imaginative modelling.[42] Aquinas's just war, as he let it unfold within an account of the virtues, did not face this shortcoming. Reichberg puts it succinctly:

> Aquinas's virtue approach has, by contrast, the advantage that it is designed specifically for such settings; thus instead of first separating reflection from practice and then facing the challenge of reuniting the former with the latter, Aquinas attempts a unified account that joins the two from the beginning.[43]

As Aquinas gave pride of place to virtue, he did not provide a detailed *jus in bello* code. That does not mean, however, that Aquinas was uninterested in how soldiers would fight. Rather, right conduct in war, for Aquinas, was linked to right intention and virtue.[44]

So what is the relevance of virtue vis-a-vis military conduct? I do not have the space here to provide a detailed account of virtue ethics. I will therefore concentrate on the aspect that seems most immediately apparent with regard to military operations, namely, the cultivation of good habits. Good habits enable an actor to consistently make good decisions. How does an actor acquire such good habits? The answer is by having rightly regulated passions. Passions themselves are properly governed when they are subject to reason.[45] In Cole's words:

> The successful integration of reason and passion allows us to abbreviate deliberation into "snap decisions" that are rational by abridgment. The virtuous person is able to act – or better, react – rightly because such "snap decisions"

are a product of a will in which the passions are integrated with reason to such a degree that acting 'passionately' is in accordance with some good end. Put differently, the virtuous person's quick reactions are rationally habitual and not instinctual.[46]

Conclusion

The crucial relevance of internalizing good habits in a military setting seems immediately apparent. Soldiers on operations must know the rules for using force. Commonly, they even carry printed cards with the rules of engagement (ROE) with them. These ROEs are derived from the four main LOAC principles. However, knowing the rules alone does not yet lead to acting well. Especially in the heat of battle, being able to act according to habit, not instinct, takes on a fundamental role in ensuring right conduct, and avoiding excessive uses of force. Not surprisingly, pushing soldiers to the threshold of where they would "snap," or lose the control about their action by giving in to the arising passions, is an important part of military training.

So how does virtue apply to military necessity? Cole takes Aquinas's emphasis of virtue and his secondary only interest in rules as a deliberate feature of his just war thinking that functions as a reminder that any rule needs to be translated into concrete action.

> The end of the *jus in bello* is to approximate virtuous behavior in battle, but the specified approximations can never be carved in stone. Put more clearly, what *jus in bello* rules strive to achieve – proportion and noncombatant immunity – is always binding upon the soldier and commander, but what is not (cannot be) binding is what may count as meeting the criterion of proportion or noncombatant immunity.[47]

Consequently, the morally right interpretation of the rules depends on the virtuousness of the actor. This aspect is of fundamental importance as far as military necessity is concerned. Above, I argued that Neff leaves out DDE when he argues that Vitoria allowed the killing of innocent persons if called for by military necessity. In other words, for Vitoria, as well as for Aquinas, in war, the principle of discrimination can be violated as a result of DDE, but it must never be violated intentionally. Indeed, intention and its inherent connection to virtue is at the very heart of DDE properly understood. The emphasis on "properly understood" is crucial, not least because of Elizabeth Anscombe's concern about the possible abuse of DDE by simply claiming that a violation of noncombatant immunity happened unintentionally.[48] Such behavior, for Aquinas, would be sinful. DDE depends on the actor's rightful intention and is thus directly connected to virtue. Consequently, in order to make the right call on military necessity, soldiers must be virtuous. Not only does virtue function as a safeguard against a morally indefensible abuse of DDE, but it also helps soldiers take morally good decisions in the heat of battle that are informed by habit, not instinct. Just as the principle of discrimination can be abused is the principle of proportionality under the threat of being taken to legitimize morally indefensible combat action. As making the proportionality calculus in battlefield situations depends on human judgment, it is always liable to human error or even deliberate abuse. Training in the military virtues, in addition to the stress on LOAC, can help soldiers act well. To conclude with Cole's eloquent summary: "The virtuous Thomistic soldier is the kind of person who knows how to act on the field of battle because he is who he is."[49]

So where does this leave us with regard to the place of military necessity in contemporary just war thinking, the question this volume seeks to answer? Should we add a distinct military necessity principle to the *jus in bello*, in addition to those of discrimination and proportionality? Eric Patterson thinks that we should. Arguing for a distinct military necessity criterion understood as stewardship, Patterson's take is that such a principle would guarantee a "richer" *jus in bello*.[50] For Patterson, military necessity as stewardship should be understood as "the careful management of that which has been entrusted," namely, human life and property.[51] Those in authority, and soldiers as their representatives, may take human life and destroy property, but they must do so in a way that is morally defensible. As the constituent elements of military necessity, he imagines the principles of troop protection, economy of force, military effectiveness, and an aim toward victory. Importantly, he advocates putting military necessity first on the list of *jus in bello* principles, before the discrimination and proportionality principles. For Patterson, military necessity as stewardship takes on a unique role, as it reacts to the practical issues troops face on the battlefield. Having military necessity in first place should not be seen as a strict hierarchy, though. Rather, he imagines a "working relationship" with discrimination and proportionality, just as the *jus ad bellum* principles should not be treated in a check-list manner. Overall, by adding military necessity as a distinct principle, Patterson seeks to establish a *jus in bello* that is more aware of the urgencies soldiers and commanders face in battle, and that does not lose sight of the inherent connection between *jus ad bellum* and *jus in bello* through the focus on victory. Patterson laments that military necessity "is almost entirely neglected" by contemporary just war thinkers and there is no consensus about the question of where the place of such a revived principle should be.[52] He points to Nigel Biggar and Keith Pavlischek who see the place of such a reintroduced principle as part of the proportionality criterion. Patterson also notes that other thinkers consider military necessity as in tension with the more important discrimination and proportionality principles. Consequently, military necessity should only follow after having considered the other two.

The contribution of the classical just war I imagine affirms Patterson's concern for the practical problems soldiers face on the battlefield. The way I understand Aquinas, he would have been very sympathetic to a just side's concerns for troop protection, economy of force, military effectiveness and an aim toward victory. While in this chapter I have given most attention to the question of when not to use military force, Aquinas clearly affirmed that just combatants are allowed to kill their opponents. The primacy he gave to the *jus ad bellum* inherently comes with the acceptance that military force may be used to bring about victory, defined as a just peace. Naturally, if the moral requirements for fighting a just war are so restrictive that they rule out victory, or at least greatly advantage the unjust side, the overall goal of any just war, namely, a just peace, will be in jeopardy. At the same time, however, also inherent in Aquinas's just war is a prohibition on a *jus in bello* code of anything goes. The arguably most important aspect of this prohibition, besides the proportionality consideration, is the discrimination principle. Consequently, the *jus in bello* concerns that Patterson seeks to capture in a distinct military necessity criterion are shared by Aquinas.

Having noted this common ground, the classical *bellum justum* can also help us grapple with the question of whether we should add a distinct moral principle of military necessity. The answer that I derive is more skeptical than Patterson about the merit of expanding the established *jus in bello*. As I pointed out above, Aquinas was not against rules per se. Rather, his emphasis on virtue sought to make sure that soldiers would naturally fight with *recta intentio*. It needs to be remembered that Aquinas, as a medieval theologian, believed in the immortality of the soul. Decisions made in this world would build up and create a soul over time that would either be oriented toward the good or disordered. This aspect of Christian theology partly explains why the moral virtues, as encapsulated in the right intention criterion, came before specific rules of conduct. The development of the canonical rules on war conduct originated from the pastoral need to counsel returning soldiers upon their return from the battlefields. The main concern in the confessional was whether they had sinned by having exhibited wrongful intentions. The concern of having broken the rules followed after that.[53] As a result, returning to the question about the need for a distinct military necessity principle, Aquinas's classical understanding of just war would not be opposed by and of itself. The main question, though, would be whether such a principle can help soldiers act with right intention.

Consequently, given the above, adding a distinct military necessity principle would be of secondary importance. St. Thomas's emphasis of virtue operates from the conviction that knowledge of the rules alone does not guarantee just conduct. Soldiers need to be virtuous to be able to make the right decisions in the heat of battle. Thus, returning to the practical aspect of how best to train our troops, an increased concern for military necessity could be built around a more detailed engagement with virtue ethics. LOAC training equips soldiers with the knowledge about what can legally be done in war. The just war teaching at military academies does not only provide future military leaders with an overview about where LOAC is coming from, but, even more importantly, why it is important to have such rules. Teaching just war also raises the awareness that *legal* conduct is not always the best *moral* conduct. At times, the morality of war can demand that combatants go beyond what is legally permissible. In order to make such very difficult decisions, especially in mortal combat, soldiers need more than just knowledge of the rules. They need to be virtuous. In this sense, the ongoing revival of virtue ethics within the field of military ethics is a laudable development. The interplay of the triad of LOAC, just war, and military virtue, I believe, points to the right understanding of military necessity.[54]

Notes

1 The author wishes to thank Darrell Cole and James Turner Johnson for very helpful comments on earlier drafts of this chapter.
2 The International Committee of the Red Cross also lists precaution and limitation as distinct principles. See International Committee of the Red Cross, *Handbook of International Rules Governing Military Operations* (Geneva: International Committee of the Red Cross, 2013), 54.
3 James Turner Johnson, *Can Modern War Be Just?* (New Haven, CT: Yale University Press, 1984), 2.
4 See the chapter by Pauline Shanks Kaurin for an overview of contemporary treatments of military necessity in just war.
5 See especially Michael Walzer, *Just and Unjust War: A Moral Argument with Historical Illustrations* (New York: Basic Books, 2015); Jeff McMahan, *Killing in War* (Oxford: Oxford University Press, 2009).
6 Ian Clark, "Taking 'Justness' Seriously in Just War: Who Are the 'Miserable Comforters' Now?" *International Affairs* 93, no. 2 (2017): 327–341, at 331.
7 Christian Nikolaus Braun, *Limited Force and the Fight for the Just War Tradition* (Washington, DC: Georgetown University Press, 2023).
8 See Eric Patterson's chapters in this volume.
9 In that sense, the revisionist critique of Walzer's moral equality of combatants thesis marks a return to the classical just war position.
10 Stephen C. Neff, *War and the Law of Nations: A General History* (Cambridge: Cambridge University Press, 2005), 64.
11 Ibid., 65.
12 Ibid., 64–65.
13 For a discussion of the Catholic understanding of military necessity since the twentieth century, see Pedro Erik Carneiro's chapter in this volume.
14 Francisco de Vitoria, cited in Gregory M. Reichberg, Henrik Syse, and Endre Begby, "Francisco de Vitoria (ca. 1492–1546): Just War in the Age of Discovery," in Gregory M. Reichberg, Henrik Syse, and Endre Begby, eds., *The Ethics of War: Classic and Contemporary Readings* (Oxford: Blackwell Publishing, 2006), 288–332, at 324.
15 Ibid., 325.
16 Thomas Aquinas, *Summa Theologica*, Trans. Fathers of the English Dominican Province (Allen, TX: Christian Classics, 1948), II–II, q. 64, a. 7.
17 Gregory M. Reichberg, *Thomas Aquinas on War and Peace* (Cambridge: Cambridge University Press, 2017), 173.
18 Ibid., 255.
19 For a detailed discussion of these attempts, see James Turner Johnson, *Just War Tradition and the Restraint of War: A Moral and Historical Inquiry* (Princeton, NJ: Princeton University Press, 1981), chap. V.
20 Ibid., 125.
21 Alex J. Bellamy, *Just Wars: From Cicero to Iraq* (Cambridge: Polity, 2006), 40.
22 Johnson, *Just War Tradition and the Restraint of War*, 143.
23 Ibid., 143–144.
24 Aquinas, *Summa Theologica*, II–II, q. 40, a. 1.
25 Johnson, *Just War Tradition and the Restraint of War*, 139.
26 See Catherine Connolly, "'Necessity Knows No Law': The Resurrection of Kriegsraison through the US Targeted Killing Programme," *Journal of Conflict & Security Law* 22, no. 3 (2017): 463–496, at 466.
27 Aquinas, *Summa Theologica*, II–II, q. 40, a. 1.
28 Johnson, *Just War Tradition and the Restraint of War*, 123.
29 See James Turner Johnson, *Morality and Contemporary Warfare* (New Haven, CT: Yale University Press, 1999), 36.

30 Aquinas, *Summa Theologica*, II–II, q. 40, a. 1.
31 See the reference and discussion at the beginning of Pauline Shanks Kaurin's contribution to this volume.
32 Reichberg, *Thomas Aquinas on War and Peace*, 79–80.
33 Darrell Cole, "Thomas Aquinas on Virtuous Warfare," *Journal of Religious Ethics* 27, no. 1 (1999): 57–80.
34 Reichberg shows how Aquinas singled out two virtues as especially relevant to the military, namely, military prudence and battlefield courage. See Reichberg, *Thomas Aquinas on War and Peace*, chaps. 4, 5.
35 Cole, "Thomas Aquinas on Virtuous Warfare," 62.
36 Ibid., 75.
37 Ibid., 76.
38 See, for example, Peter Olsthoorn, *Military Ethics and Virtues: An Interdisciplinary Approach for the 21st Century* (Abingdon: Routledge, 2011); Michael Skerker, David Whetham, and Don Carrick (eds.), *Military Virtues* (Havant: Howgate Publishing, 2019). Moreover the *Journal of Military Ethics* has been an important venue for scholarship on the military virtues. See the special issue Volume 6, Issue 4 (2007).
39 Cole, "Thomas Aquinas on Virtuous Warfare," 57.
40 Ibid., fn. 1. See National Conference of Catholic Bishops, *The Challenge of Peace: God's Promise and Our Response: A Pastoral Letter on War and Peace by the National Conference of Catholic Bishops*, 3 May, 1983. Available at: https://www.usccb.org/upload/challenge-peace-gods-promise-our-response-1983.pdf. Last accessed 10 March, 2022.
41 Ibid., 58.
42 See A. J. Coates, *The Ethics of War* (Manchester: Manchester University Press, 2016), 14.
43 Gregory M. Reichberg, "Thomas Aquinas (1224/5-1274)," in Daniel R. Brunstetter and Cian O'Driscoll, eds., *Just War Thinkers: From Cicero to the 21st Century* (Abingdon: Routledge, 2018), 50–63, at 53.
44 Cole, "Thomas Aquinas on Virtuous Warfare," 58.
45 Ibid., 61.
46 Ibid.
47 Ibid., 71.
48 See G. E. M. Anscombe, *Ethics, Religion and Politics: Volume Three: The Collected Philosophical Papers of G. E. M. Anscombe* (Oxford: Basil Blackwell, 1981), chap. 6.
49 Cole, "Thomas Aquinas on Virtuous Warfare," 77.
50 Eric Patterson, "Returning Military Necessity to the *Jus in Bello*," *Journal of Military Ethics* (forthcoming).
51 Ibid.
52 See Patterson's chapter in this volume.
53 I am grateful to James Turner Johnson for a very helpful comment on this question.
54 For an account of how this triad can inform the training at military academies, see Lonneke Peperkamp and Christian Nikolaus Braun, "Contemporary Just War Thinking and Military Education" in Tine Molendijk and Eric-Hans Kramer, eds., *Violence in Extreme Conditions: Ethical Challenges in Military Practice* (Berlin: Springer, 2023), 101–117.

4 What Is Military Necessity? A Defense of the Marginal Interpretation

*David Luban**

The Weight of Necessity

"Necessity" is a heavy word. In logic, necessary truths are those that could not be otherwise. In the material world, effects follow necessarily from causes according to natural laws that are beyond human control. In the human world, we call food, water, and shelter "necessities" because without them, we would perish. In Plato's cosmology, the divine Craftsman (the 'Demiurge') who creates the world does not write on a blank slate; even the Craftsman is limited by Necessity (*Ananke*).[1] In all these contexts, necessity means inevitability, fate, the unavoidable. A necessity claim, viewed in this light, is an assertion that something is a brute fact, whether we like it or not.

That gives appeals to necessity's enormous justificatory weight. "It could not be otherwise" is not only an assertion of brute fact; it is also a justification we offer when it seems on its face that we have done something wrong. In criminal law, the necessity defense justifies law-breaking if it is the only way to avoid a greater evil; the classic, stereotypical, example is that of campers caught in a deadly blizzard who save their lives by breaking into an unoccupied mountain cabin to take shelter. Necessity, then, becomes a normative, justificatory concept as well as a concept denoting physical inevitability. Shelter is a physical necessity for the freezing campers; without it, they will die, given the premise that trespassing is a lesser evil than death and that physical necessity generates a normative justification for otherwise-wrongful conduct.

The same is true in the special case of military necessity: When valid, military necessity justifies acts that can include extreme violence. Examples of military necessity include cases like these: It was necessary to kill enemy fighters to break the siege. It was necessary to bomb that apartment building because of rocket launchers in the courtyard. It was necessary to blow up the bridge to keep the enemy from reinforcing its forward position. It was necessary to devastate the enemy's forces to make them surrender. Claims like these can be true or false; when they are true, they justify the violence.

Military lawyers will be quick to point out that the doctrine of military necessity imposes limitations as well as justifications: It justifies some acts of violence, but it also prohibits unnecessary violence and cruelty. *The U.S. Department of Defense*

DOI: 10.4324/9781003390398-4

Law of War Manual pairs the principle of necessity with a principle it labels "Humanity," which is described as "the principle that forbids the infliction of suffering, injury, or destruction unnecessary to accomplish a legitimate military purpose."[2] The Manual further explains:

> Humanity is related to military necessity, and these principles logically complement one another. Humanity may be viewed as the logical inverse of the principle of military necessity. If certain necessary actions are justified, then certain unnecessary actions are prohibited. The Principle of Humanity is an example of how the concept of necessity can function as a limitation as well as a justification.[3]

Perhaps because they are so tightly connected, many commentators roll Necessity and Humanity into a single doctrine that they label "military necessity." This can create terminological confusion. But, regardless of terminology, the point I wish to emphasize is that in both its justificatory and prohibitive roles, necessity is a *normative* concept. To borrow Wittgenstein's familiar terminology, necessity plays a role in a language game of justification—of moral challenges to the use of violence, and moral or legal justificatory responses. This is an essentially different language game from the logician's or natural scientist's—or military strategist's—cataloguing of the brute facts of the world that Plato calls *Ananke*.

In this chapter, I explore the way that military necessity functions in moral and legal discourse about the *jus in bello*. My illustrations are various legal definitions of military necessity, but these are not solely a matter of positive legal and doctrinal argument: they are moral definitions as well.

Taking Necessity Seriously

The starting point is what I will call *taking necessity seriously*. Warfare imposes its own physical and spiritual demands, what the 1880 *Oxford Laws of War on Land* (quoting Jomini) called "inexorable necessities."[4] If the enterprise is a morally serious one, its inexorable necessities are equally serious, and the law of war cannot wish them out of existence. Whatever restraints the law imposes must accommodate themselves to military necessity and regulate around its margin.

That implies that any legal regime that would make it impossible, or even inordinately difficult, for fighters to wage war, cannot be serious.[5] If taking the law at face value guarantees defeat for one side or the other in advance, then taking the law at face value cannot be the right way to interpret it.

This is not a merely hypothetical point. To take a current important example, the existing law on human shielding, articulated in Additional Protocol I to the Geneva Conventions, arguably violates the stricture of taking military necessity seriously. On the one hand, it forbids the use of human shields (placing civilians among military forces).[6] From the point of view of guerrillas or partisans who face the overwhelming firepower of a state military, this amounts to requiring the fish to leave the water voluntarily, the better to become prey for the fishermen. On the other hand, the law requires militaries confronted by an enemy that (illegally) uses human shields to maintain the principle of distinction, which would very likely require them to refrain from attacking partisans deeply embedded in a civilian population.[7] The law may have looked like a plausible compromise to those who negotiated the treaties, but in reality, it creates requirements that neither side in an asymmetrical conflict can honor without accepting defeat.

Terrible things happen in wars. The point of the laws of war cannot be to abolish those terrible things. The point can only be to shrink them to what is necessary, where, awful as it is, necessity always means someone else's tears.[8] Any sane person wishes that war itself could somehow be made unnecessary, but if that ever happens, we will no longer need to talk about the law and morality of war. So long as war is real, the law of war must take necessity seriously.

Civilianization of the Law of War

It may seem obvious or even self-evident that only warriors can judge what is or is not militarily necessary, and therefore that military professionals should hold trumping power over what the laws of war prescribe. We must reject this idea, however. Saying that military professionals and military lawyers ought to own the laws of war is like saying that Wall Street ought to own securities regulation. Militaries aren't the only people in the battle space, or the only ones war affects. Civilians, quite literally, have skin in the game. It may be that in their origin, the laws of war were an honor code for warriors who own the battlespace by force of arms—in other words, a concession by the strong to the weak, which the weak have by grace and not by right because what matters is not the rights of the weak but the power of the strong.

But that is no longer how most people regard the laws of war. In the past decades, war makers have had to come to terms with a political world in which their actions fall under the intense scrutiny of media and public opinion—and students of Clausewitz understand that public opinion makes or breaks military causes. The international criminal tribunals give visible structure and focus to that public opinion, but it matters less that war makers might come under the ICC's jurisdiction (they usually won't) than that they come under CNN's and YouTube's jurisdiction. Their deeds will be on the Internet within minutes of engagement, captured by cell phones.

This phenomenon has become a source of frustration to military lawyers, because audiences assume that gruesome footage of dead civilians always signify war crimes even when they are nothing of the sort. The lawyers' complaints are often correct, but public scrutiny and accountability is a fact of life, and at bottom it represents a larger truth that the complaint overlooks: The civilian world has staked a claim to the way wars are conducted that is not going to go away, and that should not go away. The laws of war are now common property, as they should be.

The phenomenon I am describing might be called the *civilianization* of the laws of war, by which I mean simply that in the laws of war civilian interests matter as much as military interests. The phenomenon—Theodore Meron called it the "humanization of humanitarian law"[9]—is unmistakable.

Of course, the civilian-military schema is too simple. The people of the state fielding the army are also civilians—and they cheer for their own team, care little about the enemy's civilians, and stand firmly against holding their own army accountable for war crimes. So the civilianization of the laws of war will be uneven and inconsistent. On the other side of the ledger, it would be a disastrous mistake to assume that the military doesn't care about the human dignity of civilians, so the term "civilianization" might be a misnomer that wrongly denigrates the consciences of warriors.

I will nevertheless stay with the name "civilianization," which singles out the important fact that the laws of war have passed from exclusive ownership by militaries to joint ownership with the civilians whose fate they determine. As I now argue, the civilianization of the law and morality of war is not simply a case of wishful thinking by humanitarians who pretend that military necessity barely exists.

The reason is that, properly understood, military necessity itself conceptually requires taking civilian interests into account. Before offering that argument, it is worth previewing what it implies: that viewing military necessity as a trump card over humanitarian concerns misunderstands military necessity itself.

The Career of Military Necessity

What, exactly, is military necessity? As the following discussion demonstrates, the concept has evolved over time, and different formulations vary dramatically on key issues such as whether military necessity is a legal concept or, on the contrary, an extra-legal fact that sets the boundary on the regulation of warfare; whether "necessity" means what the literal language suggests—something unavoidable if war is to be waged successfully—or merely whatever the military finds convenient; and, above all, whether judgments of military necessity must take civilian interests into account.

Legal and Extra-Legal Necessity

In medieval just war theory, the principle of necessity "meant that the just side was permitted to use whatever degree of force was strictly necessary in the particular circumstances of the case to bring about victory. Beyond that point, *all* force became unlawful."[10] Under this conception, necessity serves both the prohibitive and the licensing functions described above. It prohibits gratuitous, wanton violence—unnecessary violence. But it licenses all non-gratuitous violence, that is, violence essential to the military goal. In Suárez's words, "if an end is permissible, the necessary means to that end are also permissible."[11]

In the medieval tradition, however, the end is permissible only if the war is just. From this, it follows that a warfighting organization waging an unjust war (in the *ad bellum* sense) can never appeal to military necessity to justify violence. The plea of military necessity is available only to the righteous. In this respect, the medieval theory closely resembles contemporary revisionist just war theory.

Medieval theory derived its just war doctrines from natural law, so it makes sense to describe both the prohibitive and licensing sides of military necessity as legal doctrines, although positivists might put scare-quotes around "legal." But starting with Hobbes another way of thinking about the relationship between law and military necessity appears. In war, Hobbes tells us, "The notions of right and wrong, justice and injustice have … no place. Where there is no common power, there is no law: where no law, no injustice."[12] Military necessity is neither a legal concept nor a moral one, because in war there are no legal or moral concepts.[13] Of course, nothing forbids states from writing rules of warfare and agreeing to enforce them either directly or through the indirect mechanism of reciprocity. The resulting laws of war are positive, not natural law, because states write them on a Hobbesian legal blank slate. And military necessity represents the limit of rational regulation—Jomini's "inexorable necessities" that combatants cannot expect each other to forego. Military necessity therefore lies outside the law, as a limiting condition on what the law can require.

The latter view represents the regulation of war promulgated in the nineteenth century. Consider the St. Petersburg Declaration of 1868, which explains its aim as follows:

> Having by common agreement fixed the technical limits at which the necessities of war ought to yield to the requirements of humanity, the Undersigned … declare as follows:…
>
> That the progress of civilization should have the effect of alleviating as much as possible the calamities of war;
>
> That the only legitimate object which States should endeavour to accomplish during war is to weaken the military forces of the enemy;
>
> That for this purpose it is sufficient to disable the greatest possible number of men;
>
> That this object would be exceeded by the employment of arms which uselessly aggravate the sufferings of disabled men, or render their death inevitable;
>
> That the employment of such arms would, therefore, be contrary to the laws of humanity.[14]

The argument proceeds from a technical analysis of the necessities of war to a state agreement to prohibit arms that exceed necessity, as "contrary to the laws of humanity." Here, plainly, "the necessities of war" lie outside the purview of the laws of humanity, and set the "technical" limits to which "the requirements of humanity" must "yield." Necessity trumps humanity.

The most famous expression of this conception of necessity as an extra-legal limit to the law is the Prussian military maxim "*Kriegsraison geht vor Kriegsmanier*": the necessities of war (*Kriegsraison*) take precedence over the rules of war. The translation loses the Enlightenment flavor of the maxim, the contrast between *raison* (reason) and *manier* (manner, style). Military necessity represents reason, which uncovers immutable scientific laws—the technical side of military science exemplified most famously by Clausewitz—whereas rules of war are merely stylistic. The immutable laws are those of physics and psychology: explanations of what it takes physically to "disable the greatest possible number of men," and of how much death and destruction the enemy must suffer to cause it to cry uncle.

Under this doctrine, the licensing function of military necessity remains grounded in natural law, but natural law means scientific laws, not moral laws. If an act of violence is indispensable to victory, military necessity permits it, no matter how destructive or cruel it is, or how innocent the target of the violence may be. If the only way to conquer a city is to besiege it and starve its population into surrender, then starving the civilians is militarily necessary. As for the prohibitive doctrine of military necessity, 'ought' implies 'can': if it is technically impossible to win the war under a given prohibition, the prohibition has no force. The only prohibition is on gratuitous cruelty—gratuitous in the objective sense that victory could be achieved with less cruelty. In effect, the doctrine of *Kriegsraison* subordinates law and morality to cold hard fact.

The Hostages Formulation

The main feature of *Kriegsraison* on which I'm focusing is that it conceives military necessity as a limit to the law that stands outside it, not a doctrine within the law. This outlook changed in the wake of World War II, where victors and vanquished alike recoiled from the seemingly limitless violence the war inflicted. The post-war formula for military necessity appeared in the second round of Nuremberg trials, in the *Hostages* case.

> Military necessity permits a belligerent, subject to the laws of war, to apply any amount and kind of force to compel the complete submission of the enemy with the least possible expenditure of time, life, and money.[15]

Call this the *Hostages formula*. The *U.S. Department of Defense Law of War Manual* follows the *Hostages* formula in its definition of the principle of necessity: "Military necessity may be defined as the principle that justifies the use of all measures needed to defeat the enemy as quickly and efficiently as possible that are not prohibited by the law of war."[16]

As law-of-war expert Yoram Dinstein emphasizes, "The key words here [in the *Hostages* formula] are 'subject to the laws of war.'"[17] Now, instead of law being subordinated to military necessity, the law comes first.[18] If a rule in the law of war provides an explicit exception for military necessity, well and good: the *Kriegsraison* principle applies. For example, Additional Protocol I (AP I) allows states to restrict the movements of relief personnel "in case of imperative military necessity."[19] But most of AP I's *in bello* rules provide no exception for military necessity, and according to the *Hostages* formula, they represent absolute limits on what warfighters may do regardless of military necessity, which no longer trumps humanity. For example, AP I states, "The civilian population as such, as well as individual civilians, shall not be the object of attack."[20] This rule carves out no exception for military necessity. In theory, at any rate, the *Hostages* formula rejects Hobbes's view of war as a law-free zone; now, law governs war everywhere, by decreeing which rules will bend to military necessity and which will not.

Momentous as this change is, the *Hostages* formula remains overwhelmingly slanted in favor of militaries, and grants them enormous latitude. Read literally, military necessity includes any lawful act that saves a dollar or a day in the pursuit of military victory. These are claims of 'military convenience,' not 'military necessity.' Conceiving necessity in this remarkably broad fashion is a far cry from (for example) the Lieber Code's definition of 'military necessity' as "those measures which are indispensable for securing the ends of the war, and which are lawful according to the modern law and usages of war."[21]

As Michael Walzer perceptively comments on the *Hostages* formula: "In fact, this is not about necessity at all; it is a way of speaking in code, or a hyperbolical way of speaking, about probability and risk."[22] It is worthwhile spelling out why. One argument for the *Hostages* formula is that anything that reduces risk to soldiers' lives in the slightest degree represents military necessity. Armies must place exceptional value on the life of each of their soldiers, and that is why steps taken to diminish risk to their soldiers' lives, even very slightly, are militarily necessary.[23]

The problem with this argument is that time and money get equal billing with human life in the *Hostages* formula, and it is impossible to justify cost-cutting and quickness as transcendent values on a par with human life. Here, the argument would have to run differently: Saving money and time might conceivably spell the difference between defeat and victory. For want of a nail, the shoe was lost, so the horse was lost, so the rider was lost, so the battle was lost, so the kingdom was lost—"and all for the want of a horseshoe nail." You can't know; therefore don't take the risk.

Labeling this line of thought an appeal to military necessity is, indeed, a hyperbolical way of speaking about remote risks. Under the *Hostages* formula, an air force can use non-precision-guided bombs rather than more discriminating precision-guided munitions merely because they are cheaper, or merely because commanders don't want to wait for a shipment of PGMs to arrive, and describe the choice as military necessity. That is because there is always a remote probability that the extra money or time might spell the difference between victory and defeat.

This way of thinking has devastating implications in interpreting treaty-based protections of civilians against the "incidental" (i.e., "collateral") damage of attacks. Those protections require militaries to take all "feasible" steps to protect civilians (AP I, art. 57(1)–(2)), and a plea of military necessity asserts that requiring the use of PGMs is infeasible. The *Hostages* formula would allow states to declare precautions that impose even trivial added risk, expense, or delay "infeasible" and therefore wholly optional. *Hostages* sells the brand-name "military necessity" at the cheapest possible price.

It may be that the *Hostages* tribunal formulated its definition of necessity in such a permissive way because the judges understood that they were in no position to set standards about how much risk and expense militaries must assume, and therefore left it to the discretion of states and commanders. Understood this way, the *Hostages* formula is not so much a redefinition of military necessity to mean military convenience as it is a doctrine of deference to military judgment about what really is militarily necessary. If so, however, it is a doctrine of nearly absolute deference, and that too is hyperbolical.

Essentially, the *Hostages* formula makes only a half-break from the doctrine of *Kriegsraison*. The distinctive feature of *Kriegsraison* is that it places the interests of the military beyond all regulation except the prohibition of gratuitous cruelty. In effect, *Kriegsraison* discounts the interests of civilians down to zero, the same as the suffering of enemy soldiers, which also counts for nothing or even less than nothing. The *Hostages* formula subordinates *Kriegsraison* to law, but simultaneously it inflates the concept of necessity to include anything the military finds helpful. Except to the extent built into explicit regulations, the interests of civilians and the suffering of the enemy are still discounted to zero.

Packaging anything the military finds helpful under the heading 'necessity' is not only dangerous but dishonest; but it flows naturally from an outlook in which interests other than military interests are simply invisible.

The Marginal View of Necessity

Necessity as Proportionality: Beit Sourik's Marginal View

Once civilian interests enter our moral deliberations—as they must—the *Hostages* formula becomes unacceptable. Instead, what counts as military necessity must be determined by weighing military importance against civilian damage. We already find this way of thinking in Vitoria:

> Care must be taken to ensure that the evil effects of the war do not outweigh the possible benefits sought by waging it. If the storming of a fortress or town garrisoned by the enemy but full of innocent inhabitants is not of great importance for eventual victory in the war, it does not seem to me permissible to kill a large number of innocent people by indiscriminate bombardment in order to defeat a small number of enemy combatants.[24]

This would be true, presumably, even if bombarding the fortress or town would save time and money in pursuit of victory, or even if it slightly diminished risk to soldier lives.

Proportionality reasoning of this sort in fact follows some law on military necessity, specifically the Israeli Supreme Court's sophisticated analysis of the separation barrier Israel constructed during the second intifada.[25] Representatives of Palestinians harmed by the barrier brought lawsuits challenging the legality of land seizures for the barrier as well as specific routing decisions. The Supreme Court found that under international humanitarian law (IHL), civilian property could be seized (with compensation paid) "to the extent that construction of the fence is a military necessity."[26] One disputed section of the fence was slated for a hill called Jebel Muktam. The Court deferred to the military commander's judgment that alternative routes proposed by the plaintiffs offered less security than the Jebel Muktam route. On the *Hostages* formula, that would have sufficed to demonstrate military necessity. Instead, the Court rejected Jebel Muktam because:

> … the security advantage reaped from the route as determined by the military commander, in comparison to the … route [proposed by the Palestinian plaintiffs], does not stand in any reasonable proportion to the injury to the local inhabitants caused by this route.[27]

For historical reasons, the Israeli Supreme Court is the only court in the world that is willing to adjudicate law-of-war issues in real time; its opinions therefore have exceptional importance in formulating customary IHL. Before examining the Israeli Supreme Court's reasoning in *Beit Sourik*, it is worth noting the most important point: the form of analysis represents a step away from *Kriegsraison* just as momentous as the taming of military necessity by law in *Hostages*. Now, warfighters are no longer the only ones whose interests matter. A tactic that better meets the army's security needs than any available alternatives is not militarily necessary if one of the alternatives is much better for civilians and only slightly worse for the army. Civilians, including the enemy's civilians, count. To see how they count, we must delve a bit deeper.

The Israeli Court analyzes necessity claims in three steps, which it labels proportionality tests.[28] First, it asks whether the chosen military tactic bears a rational relationship to its goal, and second, whether that goal might be attained by an alternative that inflicts less damage on civilians. The first is uncontroversial, while the second seems like mere common sense. (Common sense or not, it is incompatible with the *Hostages* formula, which permits "any amount and kind of force" to compel the enemy's submission, except force that IHL explicitly prohibits.)

Lastly, and crucially, the Israeli Court weighs the gains to the military against the injury to civilians in marginal terms:

> The … act is tested vis-à-vis an alternate act, whose benefit will be somewhat smaller than that of the former one. The original … act is disproportionate … if a certain reduction in the advantage gained by the original act – by employing alternate means, for example – ensures a substantial reduction in the injury caused by the administrative act.[29]

If this is so, the original act is not militarily necessary.

The idea behind this 'marginal' view of necessity is that the true military significance of an act is the gain it provides over the next-best alternative; it is 'necessary' only in the sense that it is necessary to achieve this marginal gain in military advantage.[30] Necessity judgments are inherently comparative. If the advantage over less harmful alternatives is too small, the claim of military necessity lacks normative force.

Necessary Means Necessary

Why should we accept the marginal view of military necessity? The reason follows from the normative point of appeals to military necessity: they are used to justify inflicting harms on others. To serve that purpose, they must answer the question: Why won't less damaging alternatives do? The answer must take the form: The alternatives don't accomplish X, pointing to some tangible military advantage—and X is necessary. That raises the obvious question: Necessary for what? The answer: Necessary to avert minute risks, delays, or expenses simply doesn't do the job. Or rather, it doesn't do the job unless the injuries inflicted are even slighter. Kaurin, in her chapter here, expounds on these and related concerns. All this is why, in the end, claims of military necessity are not only disguised comparative judgments (with other alternatives), but also disguised claims of proportionality.[31]

Although the Israeli Court's analysis is sophisticated, it is not eccentric. In an important sense, denying the claim of necessity to small military advantages represents a return to the term's traditional meaning: In Vitoria's words, of "great importance for eventual victory in the war," and in Lieber's words, "indispensable for securing the ends of the war." Necessary, one might say, means necessary. It is the *Hostages* formula that should be regarded as idiosyncratic.

Koskenniemi's Objection

Some might object that proportionality judgments are no more determinate than the issues they are meant to resolve—in this case, whether an act of violence is a military necessity. Criticizing *Beit Sourik*, Martti Koskenniemi complains that "if people were able to agree on what is reasonable or proportionate, no courts, or law would ever be needed."[32] In his view, talk about proportionality "tests" invokes pretensions of objectivity that lawyers and moralists cannot possibly deliver on. Proportionality begs the questions of whose interests count and how much weight to give them. Why balance only the Israeli Defense Force's interest in 'military security' against the interests of Palestinian civilians? Why not include the interests of Jewish settlers, or for that matter the justice or injustice of the entire occupation?[33] What makes the security of illegal settlements a benefit rather than a cost?[34] The *Beit Sourik* Court simply assumes there are answers to those questions—and so, in Koskenniemi's eyes, the Court "pulls itself from the quicksand of ever receding argumentative chains onto firm ground like Munchhausen, by his own hair."[35]

Koskenniemi's imagery is striking, but matters are not as dire as he believes. I said above that claims of military necessity take on their sense as part of a practice of justifying acts of violence—to borrow Wittgenstein's familiar phrase, a "language game" of challenge and response. In response to the challenge, "Why did you do this dreadful thing?" the military answers: "It was necessary to achieve our military goals." The challenge-and-response language game presupposes an audience that does not reject the military's goals *in limine*. Koskenniemi is right that that will not be every audience.[36] Notably, 'revisionist' military ethicists who deny that morality permits acts of violence by the unjust side will not be a potential audience. For example, Thomas Hurka denies all claims of military necessity to the unjust side because no act by soldiers on a side without just cause can satisfy proportionality.[37] Similarly, the audience for claims of military necessity may not include the enemy, for whom the only thing 'necessary' is that their adversary *not* achieve its goals—although even the enemy might accept necessity claims out of reciprocity, because they too use violence and make necessity claims.

But threshold unwillingness to entertain military necessity claims does not make the meaning of military necessity indeterminate; it merely takes the concept off the table. If the concept of military necessity gets meaning from its use in a language game of normative challenge and response, the assumptions built into that language game may provide the determinacy we need. Here is how:

The initial challenge "why did you do this dreadful thing?" assumes damage to outsiders (typically civilians) that resulted from the military's violence. That gives us one term of the balance: 'civilian damage.' The necessity justification, linked to a military goal, gives us the second term: 'military advantage.'

Furthermore, in the same way that the practice of challenge and response presupposes an audience willing to entertain the response, it also presupposes an audience skeptical enough of the military to raise the challenge. It assumes, in other words, a more or less neutral audience. As such, the audience gives equal weight to the lives and other interests of all sides—another assumption that helps make proportionality judgments manageable.

Bit by bit, as we flesh out the practice of normative challenge and response that gives military necessity claims their meaning, the interests to be compared become clearer. Of course Koskenniemi is right that nothing in principle limits us to just two terms—nothing except the fact that once the inquiry broadens to include indirect costs and benefits or larger issues of justice and injustice, the language game becomes unmanageably speculative. Thus, another assumption built into the challenge-and-response practice is that it focuses on local and direct rather than global or indirect costs and benefits. In *Beit Sourik*, for example, the local and direct terms are the marginal security advantage of Jebel Muktam over alternative routes versus marginal damage to civilians in these particular villages. Of course, there is a far more important question about whether the separation barrier as a whole illegally impedes Palestinian self-determination as a whole; but that is simply a different question than the military necessity of a particular route.

Necessity, Law, and Civilianization

Kriegsraison, the *Hostages* formula, and *Beit Sourik*'s marginal view represent three different analyses of military necessity. The first sees necessity as a pursuit of victory in which a commander is unbound by law and the regard for the interests of civilians. The second subordinates necessity to law, but within the limits of law, it permits the commander to discount civilian interests completely. The third requires the commander to take civilian interests into account—and, I have argued, this is the best view of military necessity. A concept of military necessity that considers only military interests does no justificatory work. This assertion is amplified by most of the authors in this volume, including by Charles and LiVecche, for example, who insist that there should be no great space between military and humanitarian interests.

These arguments are not meant as a critique of the concept of military necessity or a proposal to jettison it in the *jus in bello* or IHL. That would be silly. Military necessities are real, and law cannot wish them away.

To illustrate with an example I mentioned earlier: Article 57 of AP I requires militaries to take all "feasible" precautions to verify that their targets are legitimately military and to minimize civilian damage. Notoriously, there is no consensus on what "feasible" means. Does it include anything technologically possible, regardless of cost or risk to the attacker? Alternatively, does it exclude anything that might increase military risk, no matter how slightly? Clearly, militaries could not reasonably accept the former, and civilians in the danger zone could not reasonably accept the latter. Some reasonable Aristotelian mean must be the right answer.

Military lawyers and their counterparts (and sometimes antagonists) in the humanitarian community may reject this mode of reasoning, which aims at consilience between antagonistic absolute positions. They may regard their opponents' arguments as nothing more than lawfare, not moral or legal truth. Why not stick to the truth, and fight lawfare with lawfare? The eminent (and famously belligerent) law of war scholar Yoram Dinstein told a conference of military lawyers to do just that. They must fight back against the legal interpretations of the "human rights-niks" who "prefer a non-violent solution to every conflict" and spin the law of armed conflict in directions that might hamper militaries in their pursuit of victory. In his fighting words: "Keep poachers off the grass."[38]

The obvious difference is that analysts arguing about the interpretation of concepts like 'military necessity' and 'feasible precautions' are not pursuing hidden truths. They are not physicists hunting for the Higgs boson or mathematicians vying to prove the Hodge conjecture. They are trying to give concrete meaning to past lawmakers' and moralists' constructions, in order to impose discipline on violence when a state or groups of states go to war. The obvious danger in an adversarial competition over who owns the doctrine of military necessity is one David Kennedy highlights: When interpretation turns into lawfare, the players' trust in each other's candor inevitably erodes, so that "as we use the discourse more, we believe it less – at least when spoken by others."[39] The result (Kennedy adds) is a law of armed combat that undermines itself and casts its own legitimacy into disrepute, even in the eyes of its practitioners.

I wholeheartedly agree with this diagnosis, but not with Kennedy's cure, which is to "be wary of treating the legal issues as the focal points for our ethics and politics."[40] In place of legalism, Kennedy calls for "recapturing the human experience of responsibility for the violence of war"—accepting that "those who kill do 'decide in the exception,' … [and] as men and women, our military, political, and legal experts are, in fact, free—free from the comfortable ethical and political analytics of expertise, but not from responsibility for the havoc they unleash."[41] His argument appears to be that debates over concepts in the laws of war are irredeemably strategic. Officers and political leaders—and, for that matter, humanitarians—find it all too convenient to fob responsibility onto lawyers and the law when in fact the law is "an elaborate discourse of evasion."[42] He would substitute pure Sartrean choice for LOAC and just war theory.

But suppose there were no laws of war. Do we really believe that more responsible decisions would result, that fewer lives would be lost, or that an alternative and better vocabulary would arise for deliberation? I see no reason to think so. Without some vocabulary for deliberation, the pure experience of responsibility floats in a vacuum and goes nowhere. Like it or not, and no matter where we end up, we must start with the vocabulary we have. That is the legal vocabulary of the law of war, heavily inflected with the just war theory of past centuries. Where else could we start? In Quine's words, "We are like sailors who on the open sea must reconstruct their ship but are never able to start afresh from the bottom."[43] Warriors and civilians are stuck with each other aboard the same leaky ship. My defense of the marginal conception of military necessity is simple and straightforward: It is the only one that correctly takes both military and civilian interests into account.

Notes

* This chapter is an abridged and significantly revised version of my paper "Military Necessity and the Cultures of Military Law," *Leiden Journal of International Law* 26 (2013), 315–349.

1 Plato, Timaeus, 48a2–5 (many editions).

2 Office of General Counsel, *U.S. Dept. of Defense Law of War Manual* (June 2015, updated Dec. 2016), §2.3, p. 58.

3 Ibid., §2.3.1.1, p. 59.

4 *Oxford Laws of War on Land* (1880) (quoting Jomini), wwi.lib.byu.edu/index.php/Oxford_Laws_of_War_on_Land.

5 Some might object that if the war is unjust, making it difficult for the unjust side to wage war is exactly what we should want. Here, however, I am assuming that *in bello* rules are symmetrical between the just and unjust sides. We write laws of war to govern future conflicts, and when we write them, we can't predict what wars will be fought or which side (if either) is the just side. This is a practical reason for separating *jus ad bellum* principles from the *jus in bello*, regardless of one's theoretical views about their separability and about asymmetrical moral obligations of just and unjust warriors.

6 Protocol Additional to the Geneva Conventions of 12 August 1949, and Relating to the Protection of the Victims of International Armed Conflict (AP I), 1125 UNTS 3 (1978), art. 51(7).

7 AP I, art. 51(8).

8 On this seemingly obvious point, surprisingly often ignored, see Henry Shue, "Laws of War," in Samantha Besson and John Tasioulas, eds., *The Philosophy of International Law* (Oxford University Press, 2010), 511–516. "I think the fundamental attitude of the laws of war ... can be well captured with the contemporary pithy phrase 'shit happens.' We are dealing with war.... The purpose of the laws of war is to constrain the 'shit' when the 'shit' happens." Ibid. at 516. It is 'shit' nonetheless.

9 Theodor Meron, "The Humanization of Humanitarian Law," *American Journal of International Law* 94, no. 2 (2000): 239–278.

10 Stephen C. Neff, *War and the Law of Nations: A General History* (Cambridge: Cambridge University Press, 2005), 64.

11 Francisco Suárez, *De Triplici Virtute Theologica, Fide, Spe, et Charitate* (The Three Theological Virtues, Faith, Hope and Charity), in G. Williams et al. (trans.), *Selections from Three Works* (1944), at 840, quoted in Neff, p. 65.

12 Thomas Hobbes, *Leviathan*, Ch. 13 (1651/1668) (many editions). Hobbes was not en-
    tirely consistent in holding that war is not a law-governed activity. In *De Cive*, he wrote
    that "... in the state of nature, it is lawful for everyone, by reason of that war which is of
    all against all, to subdue and also to kill men as oft as it shall seem to conduce unto their
    good." Thomas Hobbes, "On the Citizen," in W. Molesworth, ed., *The English Works of
    Thomas Hobbes of Malmesbury* (London: J. Bohn, 1841), Vol. 2, at 113, Ch. 8, Sec. 10.
13 Neff cites Bynkershoek and Rousseau in this regard. Neff, 148.
14 Declaration Renouncing the Use, in Time of War, of Explosive Projectiles Under
    400 Grammes Weight. Saint Petersburg Declaration (1868), Introduction, https://ihl-
    databases.icrc.org/en/ihl-treaties/st-petersburg-decl-1868/declaration.
15 *US v. List* (American Military Tribunal, Nuremberg, 1948), 11 NMT 1230, 1253.
16 *Department of Defense Law of War Manual*, §2.2, p. 52.
17 Yoram Dinstein, *The Conduct of Hostilities Under the International Law of Armed Con-
    flict* (Cambridge: Cambridge University Press, 2004), 18.
18 See W. G. Downey Jr., "The Law of War and Military Necessity," *American Journal of
    International Law* 47 (1953): 251.
19 AP I, art. 71(3).
20 Ibid., art. 51(2).
21 Francis Lieber, U.S. War Department, *General Orders No. 100: Instructions for the
    Government of Armies of the United States in the Field* (Apr. 24, 1863), art. 14. It is
    worth noticing that Lieber, like the *Hostages* formula, limits military necessity to lawful
    acts. In this respect, the Lieber Code and the *Hostages* formula agree in rejecting Hob-
    besian skepticism about law as well as the doctrine of *Kriegsraison*.
22 Michael Walzer, *Just and Unjust Wars: A Moral Argument with Historical Illustrations*
    (New York: Basic Books, 1977), 144.
23 Even this is debatable. See David Luban, "Risk Taking and Force Protection," in Yitzhak
    Benbaji and Naomi Sussmann, eds., *Reading Walzer* (London: Routledge, 2014), 277–
    301. Under this theory, armies could count force protection against remote dangers as
    an absolute value regardless of how much hardship achieving minute reductions in risk
    inflicts on civilians. In Gary Solis's words, "an attacker with superior arms would be
    free to annihilate all opposition with overwhelming firepower and call any civilian casu-
    alties collateral"—collateral, that is, to risk reduction, which the *Hostages* formulation
    classifies as military necessity. Gary D. Solis, *The Law of Armed Conflict: International
    Humanitarian Law in War* (Cambridge: Cambridge University Press, 2010), 285.
24 Francisco Vitoria, "On the Law of War," in A. Pagden and J. Lawrance, eds., *Vitoria:
    Political Writings* (Cambridge: Cambridge University Press, 1991), at 315–316, Ques-
    tion 3, Art. 1, Para. 37.
25 See HCJ 2056/04 *Beit Sourik Village Council v. Gov't of Israel (Beit Sourik)* [2004];
    HCJ 7957/04 *Mara'abe v. Prime Minister of Israel (Alfei Menashe)* [2005].
26 *Beit Sourik*, ¶ 32.
27 *Beit Sourik*, ¶ 61. For discussion, see Nobuo Hayashi, "Requirements of Military Neces-
    sity in International Humanitarian Law and International Criminal Law," *Boston Uni-
    versity International Law Journal* 28 (2010): 64–68.
28 *Beit Sourik*, ¶ 41.
29 Ibid.
30 Seth Lazar, "Necessity in Self-Defence and War," *Philosophy & Public Affairs* 40, no. 1
    (2012): 3–44.
31 This point, I believe, originates in Thomas Hurka, "Proportionality in the Morality of
    War," 33 *Philosophy & Public Affairs* 34 (2005): 38. See also Hurka, "Proportionality
    and Necessity," in Larry May, ed., *War: Essays in Political Philosophy* (Cambridge:
    Cambridge University Press, 2008), 129.
32 Martti Koskenniemi, "Occupied Zone—'A Zone of Reasonableness'?," *Israel Law Re-
    view* 41 (2008): 22.
33 Ibid, 24, citing Legal Consequences of the Construction of a Wall in the Occupied
    Palestinian Territory, Advisory Opinion, 2004 I.C.J. 131, ¶ 122 (July 9), where the

International Court of Justice rejected the Israeli separation barrier in its entirety because it impedes Palestinian self-determination and favors an illegal policy of settlements.

34 Koskenniemi, "Occupied Zone—'A Zone of Reasonableness'?," 23.

35 Ibid.

36 Koskenniemi writes that the Israeli Supreme Court's posture "relies … on a certain sympathy in the audience … For an unsympathetic audience, it will appear as indeterminate wordplay and pompous self-aggrandizement." Ibid.

37 Hurka, "Proportionality in the Law of War," 45.

38 Yoram Dinstein, "Concluding Remarks: LOAC and the Attempts To Abuse or Subvert It," in R. A. Pedrozo and D. P. Wollschlaeger, eds., *International Legal Studies* 87 (2011): 488.

39 David Kennedy, *Of Law and War* (Princeton, NJ: Princeton University Press, 2006), 135.

40 Ibid., at 167.

41 Ibid., at 171.

42 Ibid., at 169.

43 W. V. O. Quine, *Word and Object* (Cambridge, MA: MIT Press, 1960), 3, paraphrasing Otto Neurath.

5 Inevitable and Indispensable

A Conceptual Approach to Necessity in War and Conflict

Louis Bujnoch

Introduction

> It is a common experience in the history of warfare that not only war but actions taken in war as military necessities are often supported at the time by a class of arguments which, after the war is over, people find are arguments to which they never should have listened.
>
> Lord Bishop of Chichester, George Bell, February 9, 1944, House of Lords[1]

The invocation of the concept of necessity is of course no strange sight in the context of war and conflict. Historically, as well as in current international affairs, military actions and whole conflicts have been described as necessary in one way or another. Aside from such 'everyday' invocations of the concept, necessity has however, as Eric Patterson makes clear in the opening chapter, remained relatively limited in its conceptual scope. It is mostly through military necessity, which is to say through a very legal reading of the concept, which for the most part is restricted to the tactical level, that the concept remains important. Military necessity is a key part of the Law of Armed Conflict (LOAC) and it forms—as David Luban points out in this collection and in his earlier work on which he builds (Luban, 2013)—the starting assumption for most lawyers working with this body of law. For the same reasons, military necessity also forms a key part of military instruction along such concepts as distinction, proportionality and non-combatant immunity, which are also enshrined in the LOAC. Thus, military necessity is conceptualization of necessity in relation to war and conflict largely restricted to tactical decision-making and the legal assessment thereof.

Beyond military necessity however, necessity has somewhat disappeared as a conceptual component in the discourse of wars and conflict and ethical questions therein. To expand briefly on what Patterson has already demonstrated, while necessity is still invoked, this usually happens without any broader conceptual clarity. On the *ad bellum* side of international law, the International Law Commission (ILC), a subsidiary body of the United Nations (UN), has undertaken efforts to not only clarify but also to codify the concept of necessity (Ohlin and May, 2016). However,

DOI: 10.4324/9781003390398-5

this is largely outside the contexts of the use of force, conflict, and war, given the general nature of the UN legal regime (e.g., see Best, 1994). While this effort is admirable, it could not create complete conceptual clarity with respect to necessity.

Similarly, the wider context of political-strategic invocations of necessity either out of prudential reason of state-type logic or out of ethical considerations as in the Just War Tradition, reveals not just conceptual ambiguity one might expect given the nature of these different knowledge communities, but a fundamental disagreement on the purpose, effect, and very nature of the concept. This chapter will examine invocations of necessity in broader, political-strategic scenarios and seeks to establish a dichotomous reading of licensing effects, and addresses the usage of necessity in international law, touching on *in bello* as well as *ad bellum* law, before turning towards invocations of necessity in the context of political-strategic arguments of the reason of state-type logic. The second part of the chapter will use first Michael Walzer's Supreme Emergency as an example of a 'broader' invocation of necessity. Building on this, the second part of the chapter will then use a dichotomous reading of necessity as inevitable or indispensable, which Walzer briefly mentions, as a theoretical reading to distinguish between different types of necessity invocations. The chapter will conclude that such a dichotomous reading can help to reduce the conceptual complexity of invocations of necessity in relation to war and conflict and serve as a framework with which the usage of the concept can be clarified and analyzed.

International Law

Necessity, today, is probably most prevalent as a concept in international law. Within this broad rubric, it mostly appears as military necessity – a key concept of the LOAC. Here, necessity serves as the primary justification for military actions in pursuit of tactical and, ultimately, strategic goals. The LOAC sees this as a negotiation between military necessity on the one side and humanitarian considerations on the other to effect a balance between both (Draper, 1973; Sandoz et al., 1987; Ohlin and May, 2016). Finding a balance between both considerations, as Luban has reminded us, is not easy at the best of times, but it is also often complicated by clashing legal cultures where military lawyers take the primacy of military necessity as the starting point of any consideration while civilian, humanitarian lawyers include the restraints enshrined in human rights. Even the application of military necessity itself, ignoring any balancing concerns against humanitarian aspects for the moment, is not as straightforward as its proponents would like it to be. Judith Gardam (2004: 70) highlights this in her discussion on the finer definition of military necessity, stating that "necessity is a relative term and requires a determination of 'necessary for what?'"

This showcases the underlying dynamic not just of military necessity, but of the general concept of necessity invoked in the context of war and conflict – necessity as a licensing but also restraining concept. While the basic assumption in LOAC is that military necessity serves to license actions in wars in aid of the pursuit of the goals of war, in practice the effects of an invocation of the concept can prove quite a bit more complicated. This base dynamic is also evident in the context of other knowledge communities than *in bello* international law, which will be addressed in due course in this chapter. Before turning to other knowledge communities and knowledge practices, however, it behooves us to first examine the other body of law within international law —*ad bellum* law.

Helping to clarify why this volume is so important, recent invocations of necessity regarding the resort to war have been few and far between compared to other historical periods. Conceptually speaking, necessity has of course lost currency with the general prohibition of the aggressive use of force in international relations (IR) and generally with the inception of the UN legal regime (Best, 1994; Neff, 2014). Moreover, the UN, through the efforts of the ILC, strived to codify *ad bellum* international law and, as part of it, also the legal conception of necessity. Here it made deliberate attempts to change previous legal conceptualizations that gave states ample scope to invoke necessity as justification for the use of force in pursuit of their interests (Neff, 2014). Roberto Ago, the chief rapporteur for the ILC on this topic, while acknowledging that the concept used to be understood and invoked in this way, made clear that the idea of a subjective 'right of states' to war as "absolute nonsense today" (ILC, 1980: paras. 8–12).

This underlines that the efforts of the ILC are intended to transform the concept of necessity, at least as far as the *ad bellum* international law context is concerned, from a concept linked to the notion of a state's right to war, to one that permits only very narrow derogations from a state's obligations. In the context of a state's right to war, which Ago is alluding to in the quote above, necessity was an almost literal concept in the sense that almost anything a state deemed necessary could be actioned – a view prevalent among German international legal scholars around the cusp of the twentieth century (e.g., Kohler, 1917). The ILC's article 25, detailing the codified version of necessity in *ad bellum* international law, sees the derogation from duties when essential interests of a state are threatened, highlighting the exceptional nature of the necessity plea, but explicitly does not link this to the notion of self-preservation (ILC, 1980; Johnstone, 2005). Thus, necessity was transformed from a broad justifying concept into one which only applies in a narrow set of circumstances and also with narrower licensing effects where invocations are permissible.

What is clear from the above is that there are considerable contrasts in the extant conceptualization of necessity in *in bello* and *ad bellum* international law. This picture is further complicated when taking into account the historical trajectories the conceptualization of necessity has taken in each respective body of law. It certainly is clear that neither the historical, nor the contemporary usage of the concept is marked in any way by clarity of purpose or meaning, though codification efforts certainly aim towards this.

Reason of State

While the contemporary international legal context does not permit for the invocation of necessity in an essentially licensing way, and rather severely restricts the invocation as well as the effect of necessity, the underlying logic still survives today. More importantly, the invocation of necessity at this political-strategic level illustrates the complexities in the understanding of the concept in terms of its underlying purpose and intended effect. The starting point for reason of state-type necessity logic is the political community, usually the state, as the arbiter of what it deems to be necessary regarding its security and survival. This communitarian perspective can be found in contemporary academic discourse in the sprawling theoretical position of IR realism. Realism centers on the state as the most important unit in international politics, regards power and capabilities as its principal currency and sees security competition and balancing between states in one form or another. Despite its focus on power and means-ends mentality, this modern theory of realism is nevertheless distinct from its reason of state predecessor (Haslam, 2003).

With respect to the concept of necessity, Friedrich Meinecke's thought stands out as one of the clearest representations of reason of state logic, which is frequently also acknowledged in realism literature (e.g., Walker, 1993; Haslam, 2003; Bell, 2008). In his principal work on the notion of reason of state, Meinecke (1924) outlines his conceptualization of necessity with respect to the state and the notion of reason of state. Meinecke's argument starts out by linking the necessity of state [*Staatsnotwendigkeit*] to reason of state [*Staatsräson*], with the former directly telling the ruler "to act in such a way as to preserve the power of the state lying in your hands" and that they "are allowed to act in this way because there is no other way to reach this goal" (Meinecke, 1924: 13).[2] Importantly, he explicitly mentions a ruler or person in charge, thereby linking his conceptualization of necessity explicitly to the practice of statecraft. Lastly, Meinecke (1924: 13) also provides the underlying understanding of necessity in this discourse in stating that said ruler is allowed to act to preserve the power of the state "because there is no other way to achieve this goal." This implies that the circumstances that compel the ruler to act – that is to say the necessity of state – work as an exception to the rule. What is necessary to preserve the state becomes legitimate by definition.

Necessity of state, in Meinecke's (1924: 6) conception, is characterized by an "exigency" [*Zwangslage*] that compels the state to adopt "defensive and fighting means" in order to protect against threats aimed at the power of the state. While he defines the necessity of state to be arising out of compulsion of circumstance, Meinecke also stipulates that it is always teleological in nature, connected to a purpose and never as an end in itself. This, he also makes clear, is not just an important aspect for the abstract conceptualization of necessity of state or reason of state but is also evident in the concrete and practical implementation of actions by political leaders. The element of the ruler who has to act on a situation of necessity of state provides the other main line of investigation into the invocation of the concept in this discourse – decisiveness. In this discourse, the understanding of necessity is implicitly linked with the ability of political leaders to arrive at decisions in order to act on the circumstances causing a necessity of state. Thus, it is not the conceptualization of necessity itself which offers any limits on what its invocation can license but the decisiveness of political leaders which acts as sole restraining factor.

Therefore, reason of state-type necessity logic is fundamentally different from the aforementioned legal conceptions of necessity. It is a purely prudential understanding of necessity where the only limits imposed on the effects of an invocation of the concept stem from the conceptualization of the state or political community and the decisiveness with which its leaders act. That is to say, there is no in-built notion of transgression which requires a 'plea' of necessity. Necessity just lies in the nature of statecraft in international politics and constitutes an inevitable feature thereof. Thus, not only is necessity conceived of in broader terms than in the narrow legal codifications of necessity, but its fundamental understanding, purpose and intended effect are also completely at odds with other conceptualizations of necessity.

Supreme Emergency

An interesting example of where the broad and narrow conceptions of necessity come together is Michael Walzer's (2006: 255–268) account of the Supreme Emergency in his *Just and Unjust Wars* (*JUW*). Walzer is of course not an advocate of reason of state-type necessity logic. On the contrary, he opens his argument in *JUW* with a rejection of the underlying realist assumptions in the first chapter of his just war account, aptly titled *Against Realism* (2006: 3). Walzer explicitly rejects the communitarianism that underpins the invocation of necessity in the realist logic to justify recourse to violence and war. He also goes to some length to refute the logic of necessity in the *in bello* context, what he dubs the War Convention, by positing that this is merely a reading of necessity of risk and probability (144ff.). Despite this rejection of necessity logic, Walzer utilizes the same in his own account of Supreme Emergency, which this section will focus on.

In his discussion of the Melian Dialogue, Walzer combines both aspects in a manner of speaking. He criticizes the logic underlying the Athenian argument, which, he contends, essentially relies just on notions of risks they were unwilling to take, but, importantly, constitutes not a course of action without alternative (5–8). Rather, he questions the underlying normative premise that the empire needed to be preserved and highlights the lack of foresight that would give credence to a necessity argument. This makes Walzer's own account of what he dubs "Supreme Emergency" later in the same book all the more surprising, given that it is susceptible to the same criticism of foresight mentioned above. Walzer's account of Supreme Emergency relies on the notion that though all conflicts and wars are emergencies, some go beyond the normal definition of these categories. He stipulates two conditions which have to be met simultaneously in order to qualify as Supreme Emergency. Firstly, the danger has to be imminent in the strictest sense of the word and secondly, and more importantly, the consequences from inaction would have to be serious to such an extent that "our deepest values and our collective survival are in imminent danger" (251–255). A situation that meets both conditions, which he sees, for example in the case of World War II, permits the transgression of the very rules Walzer sets out in his own account.

Walzer himself acknowledges that he opens himself up to criticism with this part of his theoretical position, stating that hard cases make bad law and that this is, generally speaking, of course a slippery slope. He also further qualified his position in subsequent publications, acknowledging his essentially communitarian position, but insisting that this does not necessarily negate individual human rights (Walzer, 2004). Walzer further maintains that permitting transgressions has benefits and that his account of Supreme Emergency allows for the negotiation between the "absolutism of rights" and the "radical flexibility" of extreme utilitarianism without being able to ultimately resolve both positions (Walzer, 2004: 35–40). While political communities can be the source of great immorality, this is usually done in the face of greater immorality, such as when "the very existence of a particular community" is concerned (2004: 50).

Naturally Walzer's account has attracted plenty of criticism from a variety of angles, including important critiques in this present volume, but the reason for his inclusion in this chapter is to illustrate the complexity of the concept of necessity outside of relatively tightly defined contexts and to showcase different conceptions of necessity. Walzer describes a conceptualization of necessity that is extremely licensing in nature, while at the same time trying to render the concept as restraining by tightly restricting under what circumstances the concept can be invoked in the first place. In other words, underlying the notion of Supreme Emergency is an assumption about the importance of the state that is not far from the aforementioned reason of state-type logic, in so far as the a priori assumption regarding the preservation of the political community is concerned. However, in Walzer's account, this does not lead to an unrestrained conceptualization of necessity. Rather, he seeks to restrict the very licensing effects of this reading of necessity by narrowing down when it can be invoked.

This conundrum goes to the heart of the conceptual cacophony that can be found when examining the conceptualization of necessity. While the codification of necessity in international law gives the appearance of a tightly defined concept, even within the *ad bellum* and *in bello* international law communities, there remain questions as to purpose and effect of necessity invocations. This is further compounded where political-strategic invocations of necessity are considered that are marked by conceptualizations of a completely different nature in terms of their purpose. On the whole, therefore, the concept of necessity could probably best be described as fraying at the edges.

What Is Necessity?

The chapter has so far shown that outside of the very clearly defined legal context of LOAC, say, the conceptualization of necessity and, more importantly, the way it is invoked in the context of war and conflict is fraying around the edges. Invocations can be driven by fundamentally different understandings of the concept, disagreeing on the basic premise whether it is a licensing or a restraining concept, the purpose an invocation serves and its intended effect. Often enough, the concept is defined to be licensing while attempts are made to restrict the circumstances wherein an invocation of necessity is made possible, hence trying to render it a restraining concept. This is for instance shown through the above discussion of Walzer's Supreme Emergency – a fundamentally licensing conceptualization of necessity with an attempt to restrain through a very narrow set of conditions in which Supreme Emergency actually applies.

This contrasts somewhat with pure forms of reason of state logic wherein the state, or more generally the political community and its needs are the sole driver of what is deemed necessary and any restrictions on the concept solely derive from the effectiveness of its invocation and implementation. This of course applies also to matters of security and, ultimately, of war and conflict. Though also licensing, here the conceptualization of necessity tends to assume a broader character in contrast to what the Supreme Emergency of Walzer's devising envisions. From just this single example, it emerges that there is somewhat of a conceptual cacophony regarding the invocation of necessity in relation to war and conflict. Different knowledge communities and knowledge practices can entail fundamentally different notions not only of what necessity means in this context, but also the very basic premise to what end it is invoked. Even within international law, in a knowledge community and practice otherwise known to at least strive for conceptual clarity, there is some disagreement on this point.

In turn, this raises the question as to whether the frayed nature of necessity in the context of war and conflict can be at least somewhat alleviated in order to give it some conceptual precision. In the following section, this chapter outlines a simple dichotomous reading of necessity which, while not precisely defining necessity, helps to reduce the complexity in different conceptualizations.

Inevitable or Indispensable – a Dual Reading of Necessity

Given the fundamentally different understandings of the concept of necessity, of its purpose of invocation and interned effect, some framework is required to clarify the meaning of necessity – military or broader political-strategic – and prevent further conceptual fraying. The simple dichotomous heuristic proposed below is based on an off-hand remark by Michael Walzer in *Just and Unjust Wars*, which he does not pursue further in that work. To clarify, this heuristic is not artificially creating or imposing a single understanding of necessity but is designed to cut across existing categories and provide a simplified reading of how the concept is understood and invoked. Early on in *Just and Unjust Wars*, Walzer (2006: 8) draws on the lexical definition of necessity and its dual meaning as "inevitable" and "indispensable," without developing these categories further in his own theory. However, this dual meaning, or rather each meaning on its own and set in opposition to each other, helps to simplify the conceptual cacophony outlined above.

An inevitable reading of necessity would then be in line with the conceptualization of necessity found in reason of state-type logic, which is to say it would read as the state, or any political community, doing whatever it deems necessary to protect itself as an inevitable feature of the world. As mentioned above, this notion to some extent survives today in IR realism and related theories on security and strategy. A further implication of this reading of the invocation of necessity as inevitable is the purpose and intended effect thereof. This is not to excuse specific actions, which implies an acknowledgment of transgression of a norm, rule, or law, but by its very inevitableness assumes no such transgression took place. Here, necessity is therefore not invoked to excuse, but to justify behavior and actions by the state and its representatives – specifically in relation to war and conflict.

Indispensable, on the other hand, aligns more with modern legal definitions and codifications of necessity. Here we find the notion of a plea of necessity which underlies the understanding of the concept, its purpose, and intended effect. That is to say, in this reading, necessity serves to excuse actions, implying the admission of a transgression of rules, norms, or laws. This transgression is then excused with a reference to necessity, whereby the exact understanding or codification of necessity details the circumstances thereof and the scope of license this entails. As mentioned above, this aligns with many representations of necessity in international law, either as military necessity in LOAC, or more recent codification efforts of necessity by the ILC in *ad bellum* international law.

The latter is a good example of how the understanding has shifted from inevitable to indispensable. As mentioned above, international legal doctrine used to acknowledge a right to war by states was often linked to necessity, and, confusingly, military necessity. This was of course rendered redundant with the inception of the post–World War II international legal regime and the UN system. This was then also reflected in the codification by the ILC and specifically acknowledged by the ILC rapporteurs, as the quote by Roberto Ago above demonstrates. In other words, Ago and the ILC wanted to enshrine necessity as an excusing concept into international law, which allowed the transgression of laws, here the derogation from treaty obligations, under a narrow set of circumstances. A plea to necessity could excuse certain actions because they were indispensable. But the concept would no longer serve the purpose of justifying conduct based on the inevitable nature of things, which did not acknowledge any form of transgression. Hence the concept moved, over time, from the inevitable into the indispensable column in the *ad bellum* international law context.

A different example of utilizing this dichotomy, and showcasing its own limits, is Walzer's account of Supreme Emergency. The underlying logic that some wars or conflicts are of such a special nature that they constitute a 'supreme' rather than 'normal' emergency links to the notion of inevitableness in the reading of this heuristic. While Walzer tries to portray this as a very rare occurrence, he also clearly stipulates conditions that define any future such Supreme Emergencies, implying that these are bound to occur at some point or another. On the other hand, Walzer makes clear that the license that Supreme Emergency provides implies a transgression of norms and laws even if permitted in that circumstance. Thus, he clearly links it to an indispensable reading of necessity as well. This is also showcased in the narrow set of circumstances he stipulates for when an actual Supreme Emergency occurs, similarly to the ILC's efforts to narrow down the circumstances for an invocation of necessity. The fact that Walzer's account seemingly fits both sides of this heuristic on the one hand showcases its explanatory power, teasing out different understandings, purposes, and intended effects, but on the other hand demonstrates that it cannot fully cut through the cacophonous conceptual fraying that necessity entails in the context of war and conflict.

Conclusion

This chapter has outlined different understandings of necessity in relation to war and conflict across different knowledge communities and practices. This has revealed a fraying concept, which is marked not only by different understandings thereof, but, consequently, also different purposes and intended effects linked to its invocation. The dichotomous heuristic proposed in this chapter that is centered on the dual meaning of necessity as inevitable and indispensable can help to simplify and cut across this cacophony. It is a useful tool in the reading of extant as well as historical readings and invocations of the concept. Nevertheless, as the example of Walzer's Supreme Emergency has shown, it unfortunately cannot completely dissolve the conceptualization ambiguity of necessity in all cases.

Notes

1 The quote is taken from the Lord Bishop's remarks during a debate on allied Bombing Policy during World War II. Accessed at https://hansard.parliament.uk/lords/1944-02-09/debates/d9464051-78ab-4db1-8708-33784c109ad2/BombingPolicy, retrieved on 12.09.2021.
2 All translations are my own.

References

Bell, D. (2008) "Introduction: Under an Empty Sky — Realism and Political Theory", in Duncan Bell, ed., *Political Thought and International Relations: Variations on a Realist Theme*, Oxford: Oxford University Press, pp. 1–25.

Best, G. (1994) *War and Law since 1945*, Oxford: Oxford University Press.

Draper, G.I.A.D. (1973) "Military Necessity and Humanitarian Imperatives", *RDMDG*, Vol. 12, No. 2, p. 123.

Gardam, J. (2004) *Necessity, Proportionality and the Use of Force by States*, Cambridge: Cambridge University Press.

Haslam, J. (2003) *No Virtue Like Necessity: Realist Thought in International Relations since Machiavelli*, New Haven, CT: Yale University Press.

ILC (1980) "Addendum - Eighth Report on State Responsibility by Mr. Roberto Ago", 1980, A/CN.4/31/Add.5-7.

Johnstone, I. (2005) "The Plea of 'Necessity' in International Legal Discourse: Humanitarian Intervention and Counter-Terrorism", *Columbia Journal of Transnational Law*, Vol. 43, No. 2, pp. 337–388.

Kohler, J. (1917) "Not kennt kein Gebot. Die Theorie des Notrechts und die Ereignisse unserer Zeit", *Weltwirtschaftliches Archiv*, Vol. 10, pp. 403–406.

Luban, D. (2013) "Military Necessity and the Cultures of Military Law," *Leiden Journal of International Law*, Vol. 26, pp. 315–349.

Meinecke, F. (1924) *Die Idee der Staatsräson in der neueren Geschichte*, München und Berlin: Oldenburg.

Neff, S.C. (2014) *Justice Among Nations: A History of International Law*, Cambridge, MA: Harvard University Press.

Ohlin, J.D. and May, L. (2016) *Necessity in International Law*, Oxford: Oxford University Press.

Sandoz, Y., Swinarski, C., and Zimmerman, B., eds., (1987) *Commentary on the Additional Protocols of 8 June 1977 to the Geneva Conventions of 12 August 1949*, ICRC, Geneva: Martinus Nijhoff.

Walker, R.B.J. (1993) *Inside/Outside: International Relations as Political Theory,* Cambridge: Cambridge University Press.

Walzer, M. (2004) "Emergency Ethics", reprinted in *Arguing about War*, New Haven, CT: Yale University Press.

Walzer, M. (2006) *Just and Unjust Wars: A Moral Argument with Historical Illustrations*, 4th ed., New York: Basic Books.

6 Military Necessity, Catholic Thinking, and the Great Wars

Pedro Erik Carneiro

Introduction

It is said that St. Francis of Assisi met Sultan Malik al-Kamil during the Fifth Crusade in 1219.[1] There are a number of different stories of this fascinating encounter, but most agree that al-Kamil told Francis that he did not understand why Christians were fighting Muslims. The sultan quoted Jesus, "If a man strikes thee on thy right cheek, turn the other cheek also towards him." In other words, al-Kamil was casting the Christians as aggressors, despite the previous four centuries of expansionist Muslim warfare from Arabia across the Levant, North Africa, Spain, and Persia. St. Francis answered with another passage from the same Gospel (Matthew 5): "If thy right eye is the occasion of thy falling into sin, pluck it out and cast it away from thee; better to lose one part of thy body than to have the whole cast into hell." Francis seems to have argued that Muslims sought to coerce Christians to deny their faith; and, thus, some use of force was necessary and justified to defend and save Christians from almost certain death if they did not.

Apparently, the discussions between St. Francis and the sultan were cordial and spirited and even resulted in the sultan providing greater care for Christian prisoners of war. This tale, which has been told and retold over the years with different shades of nuance, reminds us that the Catholic teaching on just war thinking is informed by both Scripture and a tradition rooted in Church teaching and pronouncements based on theological reflection on contemporary contexts.

Consequently, the Catholic Church, her popes, and theologians have often engaged, and even stood against the *Zeitgeist* of a given era. The Church can rightly be skeptical of cultural and political ideologies that violate Church teaching, such as on human dignity. Catholic just war teaching exemplifies the distrust of secular ideologies because the political theologies developed by Augustine, Aquinas, Francisco de Victoria, and Bartolomé de las Casas, to name a few, counter the *realpolitik* of their day. More recently, Catholics have expressed skepticism over the moral "necessity" of using the atomic bomb (Anscombe), the utility of the League of Nations (Pope Pius XI),[2] and the necessity of certain tactics and weapons. At the same time, Catholic intellectuals, such as in the U.S. Council of Catholic Bishops' *Challenge of Peace* (1983), scrutinized, but did not disallow, that unconventional warfare may be necessary for legitimate movements resisting imperial violence.

DOI: 10.4324/9781003390398-6

The breadth and depth of Catholic teaching on the issues surrounding military necessity is beyond a short chapter, so this chapter will serve as an introduction to some of the key issues and debates. After beginning with a biblical and just war framing, this chapter will discuss some of the key issues that have required a creative response by Catholic intellectuals. Although this chapter reports on a wide range of thinking, from Augustine to Anscombe, that does not necessarily mean that this author agrees with all of their conclusions. The goal is to demonstrate some of the contributions made by Catholic thinkers, with a focus on the debates of the early and mid-twentieth century. First, however, we must look at some of the foundations of Catholic just war thought.

Foundations of Catholic Just War Thinking

Catholic just war proponents begin with the Holy Scriptures. Below are a few passages that are generally included, or assumed, in Catholic Bible-based analysis of warfare:

1 Christ condemns those who think they could drop the Old Testament with its focus on order and justice (Matthew 5:17–19; Matthew 23:1–4; Luke 16:17; Romans 3:31; and James 2:10).
2 Do not kill the innocent man that has justice on his side (Exodus 23:7).
3 Killing is not necessarily the same thing as murder (Exodus 20:13). It is licit for public authorities to wield the sword for defense and punishment (Romans 13:1–5), but private killing is illicit (Genesis 4; Exodus 20:13; and Romans 12:17–19).
4 God ordered judgment on the people of Canaan for their rejection of His laws and to prevent them from teaching "detestable worship" to the people of God (Deuteronomy 7:1 and 20:10–18).
5 Jesus acted forcefully based on righteous indignation when clearing the Temple in Jerusalem of moneychangers and showed that righteous anger (of God, which is different from vengeful wrath, retaliation, or revenge (John 2:13–17).
6 War may include elements of divine justice (Judges 6; Exodus 8:16–19; and Matthew 24:6).
7 War is an ethical issue inside a person, not from outside a person (intent of the heart). The Bible emphasizes our motivations for action as "God looks on the heart, not the outward appearance" (1 Samuel 16:7; Mark 7:21; and Matthew 15:11–20).
8 "Love thy enemy," means that even in conflict we never are called to vindictive hate, vengeance, or revenge. (Leviticus 19:18; Proverbs 25:21–22; Exodus 23:4–5; Matthew 22:39; Luke 6:35; and Mark 12:31).
9 Jesus's directive to "turn the other cheek," is based on the circumstance or a private dispute; it is entirely right to "turn the other cheek" in a private dispute; but putting children or innocent civilians at risk of violence is an entirely different matter (Matthew 5:39 vs. John 18:19–23; Matthew 5:29–30; and John 2:13–17 and 10:31–39).

10 Christians should respect soldiers (Matthew 8:5–13 and Acts 10:1–2) and
    public authorities (Romans 13:1–7; 1 Timothy 2:1–2; and 1 Peter 2:13–17),
    although there is a difference between "what belongs to Caesar," and "what
    belongs to God" (Matthew 22:21).

In addition to Scripture, Catholics have a rich legacy of theological reflection on
government, security, peace, and the use of force. "Tradition" specifically refers
to a living heritage, facts, and teachings related to Catholicism accepted throughout the generations. In history, many Catholic thinkers, saints, martyrs, and popes
argued for war and even waging war, such as Pope John X, Pope Urban II, King
Louis IX, Joan of Arc, John of Capistrano, and Sánchez del Rio. Some Scholastic
Era thinkers challenged the Islamic religion, including John Damascene, Lorenzo
of Brindisi, and Thomas Aquinas. Even the most celebrated Catholic literary authors wrote about just war in their masterpieces, such as Dante and Cervantes.
Many of these, plus many other Catholic thinkers, contributed formally or infor-
mally to Catholic tradition over the years. But, when it comes to the philosophy
and theology of war, Augustine and Thomas Aquinas are the most crucial Catholic
scholars.

As the aim in this chapter is to deal with modern Catholic thinkers in the twentieth century, I will only briefly describe Augustine's and Aquinas's thinking on
warfare. They laid the key foundations for just war to all who came after them. Specifically, I will try to show their contributions in the debate on military necessity.

Augustine is considered a founder of Christian Just War Theory. For Augustine,
just war can be defined as: (1) a reaction to wrongdoing; (2) *Benigna asperitas*,
a violent (or, better, *forceful*) but benign reaction; (3) a need; (4) an act of love
and mercy toward the enemy; or (5) an act to achieve peace. In *The City of God*,[3]
Augustine says that God uses the "scourge of war" to correct and pulverize human
corruption; and with war, God torments even the just. Augustine argues that the
wise man will declare wars out of necessity, in the face of human injustice while
lamenting human injustices and the destruction that war brings. Often, then, the
decisions about how war is fought must take into account what we would today
call military necessity: the operational logistics and battlefield choices made to
achieve success.

The early just war framework, influenced by Cicero and which Augustine developed in his different books and letters, includes the following criteria:

1 There must be *a legitimate authority* to declare war.
2 There must be a *just cause*.
3 There must be a *right intention*.
4 Leaders must consider *proportionality of ends*, balancing the expected just redress against the total harm likely to be inflicted by the impending armed action.

Over time, later thinkers, even the revised Catechism of the Catholic Church,[4] included the criteria that war must be the *last resort* and that to wage war, there must
be a reasonable *likelihood of success*. However, it is worth noting that there are

some debates regarding these final criteria. For instance, *last resort* can be considered part of the *proportionality* principle. At the same time, *likelihood of success* is not a principle supported by Augustine, Aquinas, or even by Catholic writers who fought in battles as soldiers with different probabilities of success, such as Dante, Cervantes, or Tolkien. Suppose one argues that the criterion of *likelihood of success* prevents lives from being wasted in futile cases. In that case, one can counter that this aspect is already included in the criteria of *just cause* and *proportionality*. One must remember that Augustine argued that many wars brought bloodshed and miseries, though the "wise man will wage just wars" while lamenting "the necessity of just wars." Just war thinkers should pay attention to the words "wise" and "just" when considering what is "necessary" in fighting a war.

In "Contra Faustum," Augustine addresses the question of the evil of wars. He does not relate this to the concern of the righteous (or civilian) deaths during military conflicts. On the contrary, he argues that considering the evil of war as the death of many is *"reprehendere timidorum"* (culpable cowardice or cowardly dislike). He continues:

> What is the evil in war? Is it the death of some who will soon die in any case, that others may live in peaceful subjection? This is mere cowardly dislike, not any religious feeling. The real evils in war are love of violence, revengeful cruelty, fierce and implacable enmity, wild resistance, and the lust of power, and such like; and it is generally to punish these things, when force is required to inflict the punishment, that, in obedience to God or some lawful authority, good men undertake wars.[5]

In short, for Augustine, we must first consider the moral intentions of the belligerents. Moreover, the reason for fighting (e.g., love of violence and revengeful cruelty) directly affects how the war is fought. The principle of *military necessity*, along with other ethical principles for how war is fought, put unwelcome limits on those who lust for power and are driven by "implacable enmity."

Several centuries later, Thomas Aquinas wrote on warfare in his *Summa Theologiae*, while also writing about Islam, and the challenges of armed Islamic armies in his books *Summa Contra Gentiles* and *De Rationibus Fidei contra Saracenos*. Aquinas treated wars separately in Question 40 of Part II–II of the *Summa*. Some scholars, in addition to considering Question 40, examine Question 91 of Part I–II (on laws) and Question 64 (on murder) and Question 108 of Part II–II (on revenge). Aquinas's analysis is included in two articles in Question 40 and one in Question 64 wherein he deals with intention. Intention is an essential aspect of *military necessity*.

In two articles of Question 40, Aquinas analyzed whether it is always a sin to wage war and whether it is lawful to use snares and tricks during the war.

In terms of the ethics of going to war *(jus ad bellum)*, if we have legitimate authority, a just cause, and a right intention, going to war is not sinful; it may be entirely just. Sometimes one must act against evil to defend oneself, one's neighbor, or the common good. Aquinas mentions Augustine's idea of "benevolent severity" toward enemies and brings up the other Augustinian argument that the objective of just war must be peace. This is a form of *political necessity* that leaders must act to promote security. The question for the ethics of how war is fought *(jus in bello)* has to do with the nature of operations, tactics, and weaponry.

It is relevant to *military necessity* the debate on whether it is possible to use trickery, such as ambushes during the war. Aquinas responded by arguing that certain rights and obligations to the enemy must be observed even when at war. Aquinas is against the idea that "*inter arma enim silent lēgēs*" (for among arms, the laws are silent). It is always unlawful to say a falsehood or not to fulfill what was promised; no one should deceive the enemy like that. On the other hand, he said that one is not obligated to talk about intentions to the enemy at war, so the enemy can be misled by not knowing our intentions. Aquinas stated that the first place among the precepts of military art is occupied by the art of hiding how one will try to defeat the enemy. This is not sinful fraud, nor is it repugnant to justice.

Aquinas proposed eight articles on homicide (murder) in Question 64 Part II–II. Of interest here, we have Article 7: "Is it lawful to kill a man in self-defense?" His answer gave us the seminal *Principle of the Double Effect (PDE)* that has grounded a whole philosophical field about the intentions of acts. This is particularly important to discussions of *military necessity* and will be a crucial debating point on matters of weapons of mass destruction in the twentieth century.

PDE comprises four conditions that are relevant to debating *military necessity.* The following conditions must be satisfied before an act with both good and evil effects can be judged as permissible:

1   The act intended by the agent must be at least permissible;
2   The *good effect* of this act must follow from it at least as immediately as its *evil effect;*
3   The *evil effect* must itself not be intended; and
4   There must be a proportionate, or sufficiently serious, reason for causing the *evil effect.*

One can see how the issue of potential civilian casualties in any military action is relevant here. In any battlefield engagement, the goal is to beat one's enemy. That is a sort of operational good, but it must be done using permissible means (e.g., not using poison gas or other weapons of mass destruction). It is entirely possible that there may be a few civilian casualties, but that must not be a goal of the attack and that eventuality must be taken into account during the planning. Finally, deciding to move forward with the attack must be proportionate to threats and battlefield objectives, particularly if there is a chance of a "double effect," i.e., not just the killing of enemy soldiers but the destruction of civilian life and property.

To this point, I have provided a skeletal framework for some of the biblical, theological, and philosophical issues that have animated just war criteria and reflection. The basic just war criteria, along with *PDE*, are key contributions by Catholic thinkers and are among the useful, and contested, moral principles for twentieth-century national security ethics.

Catholic Thinkers and Wars in the Past Century

Kriegsraison and the First World War

Although the concept of *military necessity* has reared its head throughout history, it took on a militaristic edge when re-conceptualized by the Prussian military in the late nineteenth century. These German military writers argued that what they called "military necessity," the "reason of war," (*Kriegsraison*) overrode all law.[6] In other words, the laws of politics and morality did not extend to the battlefield; the ethics of war were simply win or lose, and therefore, anything goes. This terrible ethic of militarism justified the German invasion of Belgium, beginning on August 20, 1914. The German army was so reckless and destructive that this assault quickly became known as the "Rape of Belgium." Soldiers indiscriminately destroyed buildings, attacked women and children, and torched cities. Patrick Houlihan documents how German attitudes were sharply different from their neighbors. To illustrate that, Houlihan told us about the reactions of two cardinals during the First World War. Attending a papal enclave, German Cardinal Felix von Hartmann met Cardinal Désiré-Joseph Mercier from Belgium. Hartmann said to the Belgian cardinal: "I hope that we shall not speak of war" and got the following answer from Cardinal Mercier: "And I hope that we shall not speak of peace."[7] One can see how the logic of *Kriegsraison* justified the perfidy and brutality of the Nazis as well.

Ironically, at the same time that the Prussians were theorizing a compartmentalized "reason of war," Professor Francis Lieber redeveloped the U.S. Military Code, which was approved by President Abraham Lincoln in early 1863. The *Lieber Code* expounds and limits the principle of *military necessity*, in tandem with principles developed previously by Catholic thinkers such as Vitoria and Suarez. Those principles limit battlefield destruction to what is *proportionate* for localized victory and proscribe the killing of civilians and the destruction of their property (*discrimination* or *distinction*). Over time the *Lieber Code* was copied by dozens of other governments, influenced the development of the law of armed conflict (e.g., the Hague and Geneva Conventions), and is the antecedent to America's Uniform Code of Military Justice. Catholic thinkers at the Vatican and elsewhere (e.g., Henri de Lubac) reflected on these principles when confronting the injustice of the Nazis. In practice, these codes typically also reflect *PDE* described above.

We should realize that philosophical idealism and nationalistic fervor dominated the thinking of many people during the First World War. It was a period marked by a world of declining empires, political assassinations, globalization of markets, and communist revolution in Russia that almost overtook Germany and other countries as well. The currents influenced some Catholic intellectuals as well. In many countries, idealistic philosophers and theologians shaped their views on the justice of warfighting to the exigencies faced by their home countries. A notorious example is the Manifesto of the Ninety-Three[8] from October 1914, endorsed by 93 prominent German thinkers and artists, declaring their unequivocal support of Germany in the war. The manifesto was signed after the devastating "Rape of Belgium." Among the signatories were a number of Catholic theologians.

The phenomenologist philosopher Max Scheler, who spent a good part of his career converted to Catholicism, was an adamant supporter of German aggression[9] during the onset of the First World War. His two renowned professors, philosophers Rudolf Eucken and Edmund Husserl, also supported the war as necessary to the German Nation. A Protestant, Eucken, not only said that Germany was waging a just war but also a sacred war.[10] The Jewish Husserl made speeches[11] for the German government to encourage Germans to engage the battle.

In France, the renowned Jewish philosopher Henri Bergson, who said in his will that Catholicism is the complete fulfillment of Judaism and that he would have become a convert, had he not seen a wave of anti-Semitism, made speeches[12] arguing against "Prussianism" (militarism, mechanization, insatiable ambition, scientific barbarism and will perverted by pride) in Germany, encouraging French soldiers to have no fear because France would win the war.

In England, the head of the War Propaganda Bureau, Charles Masterman, convened a secret meeting of renowned English writers to discuss how they could contribute to the war effort.[13] H.G. Wells, Arthur Conan Doyle, Rudyard Kipling, and G. K. Chesterton were among them. Chesterton was a renowned Catholic thinker who converted to Catholicism in 1922. In his contribution to supporting England entering the war, Chesterton wrote "The Barbarism of Berlin" and "The Martyrdom of Belgium."

The pope at the time, Benedict XV, sought to focus the Catholic Church on charity, tried to show impartiality, and famously suggested a peace agreement in 1917, which was considered suspect by both sides of the conflict. In many ways, the Vatican seemed helpless to do much to curtail the violence of the Great War (First World War).

Many military decisions made by Germany and its allies in the Great War were considered criminal, so they would need to be justified under the principle of *Kriegsraison*, such as the use of lethal gas, submarines against merchant navies and passenger ships, killing of military prisoners, destruction of the University of Leuven Library, damage of Cathedral of Reims, economic blockades, and *Judenzählung* (census of the Jews), among others. Of course, the Western allied powers were not without their own errors and responses.

In the years following the First World War, the morality of Germany's behavior was hotly contested at home. Many felt that Germany was forced to bear an inordinate amount of war guilt, a charge leveled at Berlin but not at the other Central Powers (e.g., Austria, Hungary, and Turkey–formerly the Ottoman Empire– and Bulgaria). An example of this controversy occurred in 1921, when the German Catholic philosopher Dietrich von Hildebrand went to Paris for a conference. There, von Hildebrand was asked if he admitted that Germany was responsible for the First War. He responded that he could not answer that question because he did not fully know the historical background and because the question could have different meanings. Then he was asked what he thought about the "Rape of Belgium." Von Hildebrand responded that the invasion was an "atrocious crime." That occasion was witnessed by a Nazi-sympathetic German journalist who denounced von Hildebrand in the German press as a traitor to the fatherland.

Von Hildebrand fled Munich because of the "Beer Hall Putsch of 1923" and fled Germany in 1933 with Hitler's rise to power.[14] Von Hildebrand even met future Pope Pius XII in 1935 and criticized his approach to Nazism when the future pope showed hopes for moderate Nazism.

He was one of many who tried to assert a Catholic, Christian response to the war crimes of the time. However, Catholic voices took different sides in the debates of the 1920s, in part due to the threat of atheistic and violent Communism.

Catholics and the Second World War

A decade later, in March 1937, Pope Pius XI issued the encyclical *Mit Brennender Sorge*, condemning Nazism and racist ideology. But a few years previously, (in July 1933), the Catholic Church had signed a Concordat with Nazi Germany. The Concordat received much criticism from Catholics, inside and outside Germany, and was never respected by the Nazi government.

After Augustine, Aquinas, and Victoria, probably the most quoted Catholic philosopher in the debate on war is Elizabeth Anscombe. She made seminal philosophical contributions, despite the fact that they often are in serious conflict with historic Catholic just war thinking.

In 1939, Anscombe wrote a pamphlet called *The Justice of the Present War Examined*,[15] when she was only 20 years of age and a philosophy student. She had converted to Catholicism in the previous year. Anscombe was in favor of a policy of appeasement with Nazi Germany, in part due to what she believed to be the evil of fighting war. She tried to expound on what she thought would be the Catholic Church's view. The pamphlet had a subtitle called "A Catholic View," which was removed at the request of the Bishop of Birmingham. In her pamphlet, Anscombe argued that all of the governments involved had a certain baseline of evil inherent in them—after all—weren't they all empires? Was there really a difference between the empire and wage-slavery of the French and British empires, on the one hand, and the fascists in Germany and Italy on the other? Anscombe went on to castigate the destructive nature of war, specifically citing the idea that so-called *necessity* justified horrors on the battlefield.

Anscombe's abhorrence for violent conflict in 1939 foreshadowed her rage and disgust with the Allied decision to deploy atomic bombs at Hiroshima and Nagasaki in 1945. A little more than a decade later (in 1957), Anscombe famously challenged her employer, Oxford University, on the subject. She wrote an essay, "Mr. Truman's Degree,"[16] motivated by her opposition to Oxford's decision to award an honorary degree to the former U.S. President Harry Truman. Her opposition was based on Truman's responsibility for the use of atomic weapons. Even after full knowledge of the crimes of Nazi Germany and the Japanese Empire, she argued that the Allies went too far in asking for unconditional surrender from Hitler and Japan. Marc LiVecche addresses all this rather handedly in his later chapter. What I want to note here is that, interestingly, Anscombe asserted that pacifism is a false doctrine: she did not claim that total pacifism was moral or practicable. Instead, her enduring argument, which continues to be influential in Catholic circles, derives from *PDE*. Anscombe argued that choosing to kill the innocent to achieve strategic ends is always murder. In 1961, Anscombe published an additional article extending these arguments called "War and Murder."[17] Her argument influenced later Catholic thinking, not just on how war is fought (e.g., on *military necessity*), but also on the decision to go to war (e.g., the recent "presumption against the use of force" argument).

In contrast, the famous Catholic writer G. K. Chesterton died in 1936, but he wrote more about Nazism or Hitler than Anscombe or von Hildebrand. There are also three collections of articles by G. K. Chesterton dealing with the Second World War issues.[18] Chesterton even managed to foresee the Second World War events such as Nazi Germany's agreement with Stalin's Russia against Poland and the Anschluss of 1938.

Chesterton strongly supported the war against the Germans in the First World War. He saw England as a defender "of Europe and of sanity." Chesterton saw that the Prussianism of the Great War was continued by Hitler, who, for Chesterton, in addition to bringing the plague of Prussian militarism, brought racism, eugenics, perversion of nationalism (ultra-nationalism), and the creation of a new pagan religion that persecuted Christianity. Chesterton defended the way that England and its partners fought resolutely, but without the horrific attacks that Germany leveled on civilians (and justified as "necessary").

Christians throughout the West look back on the period of the Holocaust with revulsion and shame, although the sad fact is that many allowed, at least in the early years, that Germany's heavy-handedness was somehow a form of "necessity." In contrast, there were individuals like the Protestant theologians Martin Niemoller and Diedrich Bonhoeffer, and the Catholic martyr Franz Jagerstatter. Perhaps one of the most important voices against German atrocities at the time was philosopher Edith Stein, a student of Husserl, who became a nun in 1922, died in Auschwitz in 1942, and was declared a saint in 1998. As early as April 1933, Stein wrote a letter to Pius XI in which she said prophetically:

> For weeks, not only Jews but also thousands of faithful Catholics in Germany, and, I believe, all over the world, have been waiting and hoping for the Church of Christ to raise its voice to put a stop to this abuse of Christ's name. Is not this idolization of race and governmental power which is being pounded into the public consciousness by the radio open heresy?...
>
> We all, who are faithful children of the Church and who see the conditions in Germany with open eyes, fear the worst for the prestige of the Church, if the silence continues any longer.[19]

Stein refused to countenance anyone's argument that the Nazi state was acting on some rational principle of "necessity" in clearing out so-called enemies of the state, such as the Gypsies, Jews, and other minorities who were labeled and treated with disdain by the Nazis. Hitler's "war at home" was illicit in its rationale and evil in its activities.

Conclusion

Catholic just war thinking has provided Western civilization with a set of principles, applied contextually, that are now not just a part of our philosophical and theological landscape, but are formalized in International Humanitarian Law (IHL), the UN Charter, and the war conventions. For example, the *jus ad bellum* principles advocated by Augustine and Aquinas of legitimate authority, just cause, and right intention are the bases for legal concepts such as national sovereignty and nonintervention. Later work by Vitoria, Suarez, and others on the ethics of how war is fought provided the *jus in bello* principles of *proportionality* and *discrimination*.

Many of these writers took for granted, or explicitly adumbrated, the principle of *military necessity* as a moral principle. It is important to take every reasonable action to achieve battlefield victory. This principle, as stated by Aquinas, is particularly informed and bounded by the Double Effect, meaning that both first- and second-order effects should be assessed before a major military action occurs. If the second-order effects are particularly destructive, that may call into question the morality and utility of that action. *PDE* was the primary critique used by Elizabeth Anscombe in her criticism of the use of atomic weapons in 1945.

Finally, a robust, yet restrained definition of *military necessity* helps us see the lawlessness and destructiveness of militaristic claims to *necessity*, such as Prussian and Nazi *Kriegsraison*. As Catholic and other scholars reengage in debates on *military necessity* today, e.g., in the areas of artificial intelligence, cyberwarfare, or the new nationalisms in China and Russia, it is wise for Catholic, and other, thinkers to reconsider the morality and restraint imposed by a proper doctrine of *military necessity*.

Notes

1 See, for example: Frank M. Rega, *St. Francis of Assisi and the Conversion of the Muslims* (Charlotte: TAN Books, 2007), 64–72.

2 Pius XI, "*Ubi Arcano Dei Consilio*" (Encyclical delivered December 23, 1922), Vatican, accessed November 27, 2021, https://www.vatican.va/content/pius-xi/en/encyclicals/documents/hf_p-xi_enc_19221223_ubi-arcano-dei-consilio.html.

3 Augustine of Hippo, *The City of God*, New Advent, accessed December 2, 2021, https://www.newadvent.org/fathers/1201.htm.

4 See Paragraph 2309 of "Catechism of the Catholic Church," Vatican, accessed December 29, 2021, https://www.vatican.va/archive/ENG0015/_INDEX.HTM. It must be said that the Catechism presents the teaching of the Church without elevating the doctrinal status of those teachings beyond what they otherwise are. Just war theory is not a dogma of the Church.

5 Augustine of Hippo, "Contra Faustum," Book XXII, 74, New Advent, accessed September 25, 2021, https://www.newadvent.org/fathers/140622.htm.

6 William Gerald Downey, Jr., "The Law of War and Military Necessity," *The American Journal of International Law* 47, no. 2 (April 1953): 251–262.

7 Patrick J. Houlihan, *Catholicism and the Great War: Religion and Everyday Life in Germany and Austria-Hungary*, 1914–1922 (Cambridge: Cambridge University Press, 2015), 1.

8 See The World War I Document Archive, accessed January 21, 2022, https://wwi.lib.byu.edu/index.php/Manifesto_of_the_Ninety-Three_German_Intellectuals

9   Max Scheler wrote, for instance, *Der Genius des Krieges und der Deutsche Krieg (The Genus of War and the German War)*, in 1914.
10 Rudolf Eucken wrote, for instance, *Die sittlichen Kräfte des Krieges (The Morality of the War)*, in 1914, in support of his country waging war.
11 See, for instance, the article "Fichtes Menschheitsideal (Drei Vorlesungen) from Aufsätze und Vorträge" (Fichte's ideal of humanity—three lectures from essays and lectures), of 1917.
12 Bergson's speeches were turned into a book called *La Signification de la Guerre* (The Meaning of War), of 1915.
13 Joseph Pearce, *Wisdom and Innocence: A Life of G.K. Chesterton* (San Francisco: Ignatius Press, 2015).
14 Dietrich von Hildebrand, *My Battle Against Hitler: Defiance in the Shadow of the Third Reich*, Translators and Eds. John Henry Crosby and John F. Crosby (New York: Image, 2014).
15 G. E. M. Anscombe, "The Justice of the Present War Examined," in *Collected Philosophical Papers, Volume III, Ethics, Religion and Politics* (Oxford: Basil Blackwell, 1981), 72–81.
16 G. E. M. Anscombe, "Mr. Truman's Degree," in *Collected Philosophical Papers, Volume III: Ethics, Religion and Politics* (Oxford: Basil Blackwell, 1981), 62–71.
17 G.E.M. Anscombe, "War and Murder," in *Collected Philosophical Papers. Volume III. Ethics, Religion and Politics,* (Oxford: Basil Blackwell, 1981), 51–61.
18 G. K. Chesterton, *Avowals and Denials* (London: Methuen & Co. Ltd., 1934). Available at http://www.gkc.org.uk/gkc/books/Avowals_and_Denials.html. Also see G. K. Chesterton, *The End of the Armistice* (London: Sheed & Ward, 1940); and G. K. Chesterton, *Chesterton on War and Peace: Battling the Ideas and Movements that Led to Nazism and World War II*. Ed. Michael W. Perry (Washington, DC: Inkling Books, 2008).
19 Edith Stein, "1933, 04-12 Letter of Edith Stein to Pope Pius XI", Center for Dialogue and Prayer in Oświęcim, accessed December 30, 2021, https://cdim.pl/1933-04-12-letter-of-edith-stein-to-pope-pius-xi,1798.

7 Necessity, Convenience, and Point of View

Military Necessity in Just War

Pauline Shanks Kaurin[1]

> Military necessity is limited by the principle of humanity. The Lieber Code therefore states in article 16 that: "Military necessity **does not admit of cruelty**—that is, the infliction of suffering for the sake of suffering or for revenge, nor of maiming or wounding except in fight, nor of torture to extort confessions."[2]

The above statement is a classic starting point for discussions about how requirements of morality, and just war thinking in particular, intersect with considerations for military necessity (MN) to achieve the military ends in war. Building on this starting point, it is important to address three possible common-sense views of MN before getting into the scholarly debates, critiques of those views, and then my own argument on MN.

The first view holds that MN is at odds, or in tension, with these moral requirements, such that MN may trump or limit the moral principles and how they can be put into effect. This view would mean that moral principles ought to take a back seat or become a secondary consideration to the *primary* consideration of military victory and/or success; morality is fine if it does not jeopardize the military mission. Other variations on this view consider either the moral imperative of victory taking precedence over any other moral considerations, or that the practical imperative of victory takes precedence over moral considerations. In short, military victory must come first in this view.

Another approach views moral principles as circumscribing what can be done from a military point of view of just war thinking (or other moral principles or norms), defining the outer bounds of what can be done to achieve military success/victory. This moral framing provides the structure and boundaries within which military judgments and the concept of MN must take place. In this case, there are important concerns about how this boundary is reinforced and maintained, as well as what the sanctions are for military judgments or practices that exceed its bounds.

A third position suggests that neither moral nor military considerations have complete primacy, but that one must consider them both in dialogue, with each defining and limiting the other iteratively. Military victory here is in fact a moral requirement, while at the same time just war thinking criteria, including *reasonable chance of success, proportionality* of means, *discrimination*/distinction, and

DOI: 10.4324/9781003390398-7

moral commitments to restoring peace and avoiding unnecessary suffering for combatants and non-combatants alike, apply to how one conducts all military operations. A model of negotiation might be helpful here, where the military realities and capacities are considered in light of these moral requirements; but these moral requirements also need to be seen through the lens of what is practically necessary for military victory and achieving the end of the war.

Regardless of which view of MN one takes, it is clear that a deeper treatment of the nature of MN as well as consideration for whether and how it ought to have status as a distinct moral requirement within the context of just war thinking is in order. Accordingly, this chapter presents and discusses viewpoints and considerations for MN in three sections, including: (1) the nature of MN and how MN has been defined within just war thinking parameters; (2) an examination of MN as a separate just war criterion; and (3) considerations for moving the conversation forward including an argument that any MN requirement must be conceptually distinct from *proportionality* judgments; and that judgments about MN must not only focus on a viewpoint of military culture, preference, and procedures but also consider the impact of MN harms from the point of view of those who would be harmed, using methodologies appropriate to their experiences and concerns.

What Is Military Necessity (MN)?

To begin, the scholarly treatments of MN by Michael Walzer, Helen Frowe, and David Luban will be considered, setting up some of the basic definitions and arguments in the context of just war thinking. While these are not the only views that one might consider, they are representative of just war thinking approaches and sets of concerns. Generally, MN is the view that only those actions that are required to bring about the relevant military objectives are permissible within the moral principles of just war thinking. Any actions beyond MN within just war thinking are morally problematic in that they may inflict suffering beyond what is necessary to achieve the military objective and this unnecessary suffering is therefore morally impermissible. Of course, what matters is exactly what 'necessary' means here. A strict, logical *necessity* is that quality or element which without a particular part cannot ever become the entire concept or action. For instance, if oxygen is necessary for humans to be able to breathe, that means that without oxygen, a human being cannot breathe and therefore is not a living human being. Are discussions of MN therefore really about *necessity* in this sense, or are they about something else?

Walzer's account of MN in *Just and Unjust Wars* contains several discussions worth considering for setting a basic framework for some of the other accounts of MN. For instance, Walzer identifies one such definition *(MN1)* with the German *kriegsraison* (reason of war): "The doctrine justifies not only whatever is necessary to win the war, but also whatever is necessary to reduce the risks of losing, or simply to reduce losses or likelihood of losses in the course of the war."[3] Walzer also notes that this is not *necessity* proper, for which he states "there is no right to commit crimes to shorten a war,"[4] but his assessment of *MN1* rather is that it is a matter of probability and risk.[5] Walzer further notes that MN cannot justify killing persons who are not already liable to be killed; that is it cannot justify the intentional targeting of non-combatants, and therefore Walzer's discussion of Doctrine of Double Effect and collateral damage also fits under the *MN1* model, which Walzer ultimately rejects.[6] *MN1* therefore is primarily a calculation of the reduction of loss to increase the probability of *success* relative to combatants.

The second discussion of *necessity* (*MN2*) occurs in Walzer's treatment of aggression and neutrality in the violation of Belgium in World War I.[7] Walzer notes two versions or understandings of *necessity* in the German approach to MN: (1) *necessity* as instrumental at the strategic level, which really meant improving the odds of German victory and (2) the moral argument that winning itself is necessary to the continuation of the German political community.[8] Walzer also considers Churchill's claims about the *necessity* of the violation of Norway's neutrality in 1940 relative to the existential threat posed by Nazism in terms of these same two arguments.[9] Walzer rejects the claims in *MN1* and *MN2* in the strict sense (necessary for victory; and victory as necessary for survival). He finds the *moral* argument from the *necessity* of the continuation of the political community persuasive, but not for MN, since in 1940 what was at stake was a quick victory as opposed to victory itself.

The third (and fourth discussion) of MN *(MN3/4)* is the Supreme Emergency argument and the accountability piece that follows it, as a case where MN in the truest sense (the necessity of the existential continuation of the state as protector of the rights of its citizens and the common life); in his view, the War Convention can be violated if the violation is acknowledged as immoral and punished afterward.[10] This is a temporary violation of the War Convention, as opposed to the moral justification of otherwise prohibited activities on the grounds of MN.[11] Therefore, on Walzer's account there is really only one example of MN in a true sense and it's pretty catastrophic. The rest of the discussion is more about the probability of success, the risks involved, increasing the likelihood of success, and to what degree these considerations should be a counterweight to the other moral considerations in the War Convention.

Second, there is Revisionist Helen Frowe's account which first looks at *necessity* in the context of self-defense (because Revisionists such as Frowe approach just war discussion through the self-defense lens) and then turn to MN in an offensive context. Beginning with the self-defense context, she highlights David Rodin's argument that *necessity* is only fulfilled when an attack is clearly imminent, a nod to the just war thinking criteria of Last Resort as we cannot know this until other means have been exhausted and the attack cannot be avoided.[12] In addition, this notion of *necessity* forbids the use of more force than is necessary to protect oneself, with any force beyond that being the domain of excessive force and imposing unnecessary and unjustified suffering.[13] Turning to MN *(MN5)*, Frowe notes that the offensive action must confer some advantages and be lawful, which requires some thinking about what will improve the chances of defeating the enemy, but this is seen primarily as a military judgment that ought to leave politics out of these judgments.[14] This is quite similar to Walzer's *MN1* and *MN2* formulations, which is not *necessity* in a strict sense at all; further, she seems to want to cordon off military judgment as something separate from strategic and political judgment, suggesting that she sees MN primarily in terms of a *jus in bello*, tactical level concerns.

David Luban's critiques of these and other standard views on MN center on two main issues. First, as when he insists that "necessary means *necessary*" and not merely Necessary means averting "minute risks, delays, or expenses" Luban is concerned that what is termed MN doesn't become shorthand for whatever is merely convenient or preferred for the military in terms of their traditions, norms, culture, standard operating procedures, and bureaucratic requirements.[15] Luban describes this mode as military culture, and this especially seems to be the current understanding with collateral damage and Doctrine of Double Effect. This, of course, does not mean that what is described in terms of *necessity* here is objectively necessary for military success. Rather it reflects a certain deference, largely uncritical acquiescence, to military culture and their professional judgments.[16] Luban further notes that because of this deference, there are huge incentives to lie or at least rationalize and self-deceive in response to outside judgment, including problems like moral disengagement.[17] It is largely uncritical (at least for the United States) because of the legacy of Vietnam and due to ideas of autonomy and expertise that are part of the military Profession. This ends up conflating MN with what the military finds helpful and preferential for them.

Second, Luban argues that one must take into account civilian interest, echoing the views of Rosemary Kellison and Jessica Wolfendale—and the humanitarian concerns stressed by Charles, LiVecche, and Patterson, among others, in this volume—who point out that war must be seen through the civilian experience of those directly impacted, not just seen through the military lens.[18] The issue here is about how civilians, lawyers, and the military culture all think about *necessity* and whether *necessity* simply devolves into what Walzer described as risk and probability, or even more worrisome, whether military culture sees *necessity* as what they prefer to do.[19] If Luban is right, it seems most of the time "necessary" = efficiency in consonance with our own preferred doctrine or cultural way of doing things, as opposed to what is objectively necessary for military success. Luban notes the difference between how commanders might think of this and how lawyers view this, with the lawyers thinking they ought to make the call and commanders thinking that they should, with neither party engaging the civilian interest in a full empathetic and nuanced way.[20]

At this point, there are several points to note. First, if MN is whatever the military wants/decides, this means that MN is meaningless to restrain its judgment and actions, although it may apply as a public relations move or moral window dressing to make the larger society feel better about actions that they might normally condemn or reject as immoral. As the Frenchs argue in their chapter here, given the psychological trauma, moral injury, and the undermining of healthy transitions and readiness to which violating moral norms subjects our military personnel, such moral window dressing can be dangerous. If, second, given the picture of MN so far, one might wonder what happens if their judgment is wrong, self-serving, or too narrowly focused on tactics and not enough on strategic and/or moral considerations? One might also wonder about *jus ad vim* where judgments rooted in the orientation and practice of kinetic violence and force may not transfer over to actions short of war.[21] Walzer has contrasted *jus ad vim*, as "'measures short of war' such as imposing no-fly zones, pinpoint air/missile strikes and CIA operations, and ...'actual warfare', typified by a ground invasion or a large-scale bombing campaign."[22]

Third, one might be concerned about whether MN avoids, or is even intends to avoid, military excess. This concern seems especially acute if we are thinking of MN either from the moral (and not merely legal) perspective of either *jus in bello* or *jus ad vim* because of the issues of expertise and autonomy within the culture of military professionalism where others are expected to defer to military judgment. This is in tension with the idea that the military is subject to civilian authority and is also making moral judgments that are and ought to be accessible to a broader public; however, the military still has professional moral expertise, but without a monopoly. Who ought to make moral judgments about what counts as excess, on what grounds, and in what domains? Seth Lazar notes that these *necessity* judgments are inherently comparative, so the question becomes necessary for what end?[23]

Why Should We Consider Military Necessity as a Just War Thinking Criterion?

The prior discussion highlights some of the basic concepts and questions related to the military necessity (MN) discussion in just war thinking. These complexities lead some (including some in this volume) to suggest that MN ought to be a separate and distinct just war criterion to ensure that it functions as a moral restriction and limitation or provides the impetus to cut further future losses by ending the war quickly. One might also wonder whether or to what extent MN actually provides any restraint in its current incarnation, and whether it instead serves to rationalize a current pragmatic permission structure from the standpoint of the military, failing to consider the experience, needs, and perspectives of those who would be harmed. If it functions at all, it seems that it functions more as a legal consideration, even though Lieber's formulation of the concept noted that it could only apply within what was already legal and permissible in war. A reasonable conclusion is that it does not restrict military force, but rather functions to give explanations and rationalizations of how military force is applied.

Given the discussion so far, certain constraints on any MN criteria emerge. First, any MN criteria would have to take the stance argued by several authors who consider the judgment of MN as informed by the perspective of those who will be harmed. We have discussed already how it could easily be seen as a tactical/*jus in bello* judgment, an operational judgment as a part of a conventional campaign, or a *jus ad vim* assessment, or more broadly as a strategic or *jus ad bellum/jus post bellum* judgment about certain ends as well as weapons, tactics, targets, or campaigns. If it is to truly be a moral criterion that restrains behavior, something more than rationalization/explanation that is from the military effectiveness point of view is required. MN would need to not merely consider the point of view of those harmed, it would have to privilege that view to the extent that MN could only be overridden in a case of (1) a true moral and philosophical necessity, that is, there is no alternative; (2) the benefit accrued to those harmed, particularly in the long term. In short, MN understood as military culture, efficiency, or preference involves pragmatic, but not always moral considerations; and while important, ought to be weighted as such.

Second, MN would have to look at the minimum harm threshold, as opposed to the excessive harm idea. Avoiding excessive harm is *different* from using the minimum harm necessary; one could avoid excessive, disproportionate harm and still use more than the minimum harm necessary to achieve the military objective. This is another area where a focus on the experience and impacts (both short- and long-term) of those who would be harmed is necessary; this would require a much more qualitive, narrative-oriented approach than the current quantitatively consequentialist approach. This means that we would have to reassess how one does this assessment in advance (especially in collateral damage calculations) Otherwise, it will be hard to know what the impact will be on those who would be harmed, aside from loss of life and tangible property destruction. These are, as Rosemary Kellison documents well in her book, only the tip of the harm assessment iceberg and are really assessed from the military perspective with its priorities for tangible, measurable harm.[24]

Third, both just war thinking and International Humanitarian Law (IHL) include a prohibition against unnecessary suffering and engaging in any activity that would inhibit the restoration of the peace. Any MN criteria should require that the burden be on the military to not just show they are not inflicting unnecessary (as opposed to excessive) harm, but also that their actions will not reasonably interfere with the restoration of the peace understood through the lens of *jus post bellum*. These aspects would also lend themselves to requiring more qualitative, narrative approaches and understanding of the harm inflicted from the point of view and experience of those who would be harmed. There will be some tension with military culture and practices, but this tension is necessary for there to be true restraint and/ or assurance of the quelling of all hostilities.

Finally, MN criteria would need to seriously address the issue of quick/decisive victory, which seems at play in Walzer's discussions of *MN1* and *MN2* as a part of the MN criteria. There is certainly a utilitarian argument to be made for quick, decisive victory both in terms of lessening overall combatant and non-combatant harm. In addition, there is an obligation not to waste combatant lives and other resources in war, as well as a concern that the longer a conflict endures, the more likely that worse harms will occur (not only death) for non-combatants as well, including making the restoration of the peace more difficult in *jus post bellum* considerations. There are also relational obligations to non-combatants under Ethics of Care in both the short- and long-term considerations. However, there is a danger that if quick victory is part of the MN criteria, the other considerations might be overpowered by the utilitarian consideration if it does not take into consideration the ability to restore peace. If this occurs, then we are brought back to a point where military preference and culture call the shots and reduce any meaningful restraint that the MN criteria might provide. Quick victory seems to privilege more force, not less, as well as escalation of force being judged to have a greater impact, which is not necessarily true. For all these reasons, including quick victory as part of the criteria could be deeply problematic without considering the point-of-view of those who would be harmed as well as the ability to restore peace postwar.

Aside from the question what the criteria for MN ought to consist of, another consideration is the role that external moral forces, like that of the military profession (referred to in this chapter as "the Profession"), play in moral restraint. This consideration might be as a replacement for the MN criteria or as an additional buttress and support. Yishal Beer thinks of the norms of the Profession as a restricting factor. Beer argues that if MN were subject to the Profession's norms, it would be more in line with the humanitarian formulation of MN that Luban seems to be after.[25] One core question is whether the use of unnecessary or excessive force in fact violates the Profession's norms and if so, how? Beer uses Lieber Code Article 16 (quoted at the beginning of this chapter) as evidence that the Profession does not include excessive force.

A new version of MN needs military professionalism content/constraint from a military perspective, but it is not clear that the concurrent conceptions (especially those rooted in a *warrior ethos*) of the Profession do this. The warrior ethos upholds the Profession's image of the warrior as one who is a "member of a specially set aside (with rituals that attend this) class or caste … trained in killing and combat, usually in the defense of their society and who will engage against other warriors."[26]

One might question if the *warrior ethos* is likely to result in more restraint or be effective as a limiting mechanism, in part because the *warrior ethos* subculture within some militaries might actually undermine this commitment to restraint in practice even if there is in principle agreement rooted in the Profession.[27] This can also be related to the function of the Profession and the ways in which the identity issues in the Profession drive norms, values, and commitments (moral as well as non-moral.) While many would still think of the Profession in terms of the warrior archetype/ethos, I think Dean-Peter Baker's arguments relative to my *guardian conception* are important and would, in fact, shift the nature of the arguments given by Beer in a way that can address at least some of the objections raised to MN.[28] The *guardian ethos*, as contrasted with the *warrior ethos*,

> could still be viewed as a separate class with specific training and a specific vocation or role within society, but they are servants of society and subject to its constraints in certain ways that relate to their role as protector and up-holder of the Common Good. For Guardians, violence is merely one of many means that might be used to achieve their ends, and does not define them as a group or class; violence is not integral to their existential identity in the same way it is for warriors.[29]

However, even if the *guardian ethos* mitigates some of the issues, this approach is still looking at this through the military culture lens and especially through the military's current practices. Therefore, relying on these kinds of external moral forces might seem to provide some help, but they also contribute to the problem by continuing to view the problem through the lens of the military culture.

Another issue in Beer's account is that the current conception of MN as a *jus in bello* issue does not require the attacker to explain the MN of that particular target (except to justify collateral damage). Effective contribution to military action is required for targeting objects, but not for combatants.[30] Combatants *sui generis* are legitimate targets, but Beer asks what is the personal level of threat that individuals possess? In his view, combatants are to be targeted to the degree that it is required by MN; unlike Revisionists' consideration of the individual context,[31] this would be a unit-based judgment.[32] Is targeting a particular unit militarily necessary to the end in question?

Beer's account seems to suggest that there is a certain aspect of Moral Equality of Combatants (MEC) that is relevant to MN, not just questions of non-combatant harm. Given the Revisionist critiques and rejection of MEC, one ought to consider what understanding of MEC is working here. The version of MEC advocated by Michael Skerker (against the Revisionists) seems amenable to Beer's ideas; this account is not morally equal but morally equal in justification/permission to be considered a certain way not based upon actions/threat but upon role in defense of the society and whether that role is justified.[33] This is different from the Revisionist focus on individual actions in combat, which could on occasion be an issue of individual or collective disobedience or less than obedience based upon context. Skerker undercuts the Revisionist view that denies MEC because of their focus on individual orientations/status of combatants and/or noncombatants based upon whether they are fighting in a defensive basis or could be considered unjust enemy aggressors because their arguments could be considered as going against the War Conventions that combatants are required to follow.[34]

Accordingly, the question from an MN standpoint would be whether or to what extent the members of the Profession are acting in accordance with the moral norms and requirements of that community of practice, rather than the question of individual action extracted from the collective context in which it happens, is given meaning and most importantly is given moral status and permission. Waving my hand in one context is simply not the same as waving a hand in another. Hailing a taxi and engaging in interpretative dance or signaling for help or trying to intimidate and scare are all different things and have different moral status depending upon context and community. It is not just the action of the individual absent other considerations. Surely all of this will matter for judging the MN of actions, sets of actions such as operations, as well as campaigns.

Considerations Moving Forward

Given the above discussion, where does this leave MN as a just war criterion? MN as a separate criterion would need to: (1) address the would-be victim viewpoint and experience and change the burden of proof in the discussion; (2) assess MN as separate from *proportionality* discussions to get at excessive harm and interfering with restoration of the peace in explicit ways, as it is not clear that *proportionality* gets at either of these directly. Some actions might be permitted using *proportionality* assessments that ought to be prohibited within a new MN criterion.

There are also five further concerns that would need to be addressed in any future scholarship on MN criteria within just war thinking. First, regardless of what level and scope we are considering for the criteria, it would be critical to have a clear account of how MN would be necessary and different from *proportionality* or related criteria that take *proportionality* as part of their assessment, such as risk of escalation. The intuitive counterargument to MN criteria is that it would duplicate the judgments made by various *proportionality* assessments in just war thinking, and thus is unnecessary. It may be that this criterion becomes more of a model of *reasonable chance of success* relative to risk and probability where speed of victory matters in both practical and moral terms. Of course, the concern here is going to be whether there is *any* limit on the means that can be and should be legitimated by the quick victory aspect of MN.

Second, in thinking about the advantage of MN requirements, it is worth returning to the notion in the introduction of this chapter of negotiation between moral requirements of just war thinking and military reality. If MN is the conversation and dialogue between the two modes of judgment and assessment, this dialogue cannot only take into account the military perspective, but also the perspectives and experiences of those who stand to be harmed by such action in a robust and serious way by also involving empathy, narrative, and qualitative models and other means that will accurately reflect the point of view of the would-be victims' lived experience.

Third, while the Profession as a moral restraint with a guardian model is arguably better than a warrior model/ethos, that shift does not effectively address the military culture bias and preference, even if it makes it more likely that such a view can incorporate Ethics of Care considerations from the perspective of those who would be harmed by military actions and conflict more generally. It is further necessary to address *how* these judgments are assessed, not just *who* makes the judgments so we can include narrative and other qualitative methodologies and measures of harm.

Fourth, any MN formulation would require discussions of MN in both the *jus ad bellum* and *jus in bello* contexts, at both the *strategic* and *tactical* levels, since there may be different judgments and considerations that are important in each of these contexts. In addition, there would need to be consideration of elements of diminishing the risk to one's own troops and using force short of war (*jus ad vim*) and alternative contexts that bridge different levels like James Dubik's justice of war waging (inbetween the classic *jus in bello* and *jus ad bellum* levels.)[35]

Finally, David Rodin's discussion of last resort could be helpful in thinking through MN as a separate concept since it seems to be an additional connection and possible constraint, especially in connection with *proportionality* of force (means) considerations. Given the American military preference for use of overwhelming force, Rodin's view could provide some potential for additional restraint to check this cultural preference. Last resort and likelihood of escalation (which are the two criteria that get picked up in the *jus ad vim* debates) also might give some insights and perspective here to buttress MN as a restraint on military force.

Notes

1 Disclaimer: Personal views only. Does not reflect the official position of Department of the Navy or US Naval War College.
2 See "The 'Lieber Code' – The First Modern Codification of the Laws of War" at https://blogs.loc.gov/law/2018/04/the-lieber-code-the-first-modern-codification-of-the-laws-of-war/#:~:text=Military%20necessity%20is%20limited%20by,of%20torture%20to%20extort%20confessions
3 Michael Walzer, *Just and Unjust Wars: An Argument with Historical Illustrations* (New York: Basic Books, 1977), 144.
4 Walzer, 210. Also see his discussion on MN on pp. 128–129.
5 See Walzer's discussion on "Supreme emergency" on pp. 251–252.
6 See Chapter 9 in Walzer's *Just and Unjust Wars* on the Doctrine of Double Effect.
7 Walzer, 240.
8 Ibid., 241.
9 Ibid., 249.
10 Ibid., 251 ff.
11 For example, the usual utilitarian argument for the dropping of the atomic bombs on Hiroshima and Nagasaki in World War II is that it was moral because of the consequences (ending the war sooner, with fewer casualties, etc.)
12 Helen Frowe, *The Ethics of War and Peace: An Introduction.* (New York: Routledge, 2011), 76. Similarly, in Chapter 12 Marc LiVecche offers a more nuanced and complicated moral justification of the atomic bombings.
13 Frowe, 11.
14 Ibid., 106–107.
15 Luban stresses this concern in his chapter in this volume. He elaborates on it further in his original essay: "Military Necessity and the Cultures of Military Law," *Leiden Journal of International Law* 26, no. 2 (June 2013): 315–349.
16 For more discussion on deference in military culture, see Pauline Shanks Kaurin, *On Obedience: Differing Philosophies for Military, Citizenry and Community* (Annapolis, MD: US Naval Institute Press, 2020).
17 Luban, "Military Necessity and the Cultures of Military Law," 334. On moral disengagement see Albert Bandura, *Moral Disengagement* (New York: Worth Publishers, 2016).
18 Luban, 339. See also Rosemary Kellison, *Expanding Responsibility for the Just War: A Feminist Critique* (New York: Cambridge University Press, 2019).
19 Luban, 332, 343.
20 Ibid., 326. Also see the 2015 film *Eye in the Sky* for an interesting illustration of this tension, as well as a reflection of how difficult it is for either of these groups to fully and empathetically consider the impacts on those who will be harmed from their point of view and experience.
21 For more on *jus ad vim*, see Jai Galliott, ed., *Force Short of War in Modern Conflict: Jus ad Vim* (Edinburgh: Edinburg University Press, 2019).
22 Walzer is quoted in Daniel Brunstetter and Megan Braun, "*Jus ad bellum* to *Jus ad vim:* Recalibrating Our Understanding of the Moral Use of Force," *Ethics & International Affairs* 27, no. 1 (2013): 87.
23 Lazar is quoted in Luban, p. 345. See Seth Lazar, "Necessity in Self-Defense and War," *Philosophy and Public Affairs* 40, no. 1 (2012): 3–44.
24 See Kellison, *Expanding Responsibility for the Just War.*
25 Yishal Beer, "Humanity Considerations Cannot Reduce War's Hazard Alone: Revitalizing the Concept of Military Necessity," *European Journal of International Law* 26, no. 4 (November 2015): 801–828.
26 Pauline Shanks Kaurin, "Warrior, Citizen Soldier or Guardian: Thoughts After a Kerfuffle." (blog). (December 20, 2016), https://shankskaurin.wordpress.com/2016/12/20/warrior-citizen-solider-or-guardian-thoughts-after-a-kerfuffle/. Also see Kaurin's book,

The Warrior, Military Ethics and Contemporary Warfare: Achilles Goes Asymmetrical (Military and Defence Ethics) (London and New York: Routledge, 2016).

27 The recent Brereton Report from Australia and the Chief Eddie Gallagher case in the United States both raise issues with the warrior ethos and the Profession in regard to special and elite forces and their ethical violations.

28 Dean-Peter Baker, *Morality and Ethics at War: Bridging the Gaps between the Soldier and State* (London: Bloomsbury Academic Press, 2020).

29 Kaurin, "Warrior, Citizen Soldier or Guardian: Thoughts after a Kerfuffle."

30 Article 5.2 in Additional Protocol I.

31 For more on the Revisionists' consideration of harms to individuals versus collective harms in MN decisions, see Seth Lazar, "Just War Theory: Revisionists Versus Traditionalists," *Annual Review of Political Science* 20 (May 2017): 37–54; also see James Turner Johnson, "Just War, as It Was and Is," *First Things* 149 (January 2005) on the restraints on commanders to consult lawyers in making battlefield decisions and the constraints this also places on individual soldiers.

32 Beer, "Humanity Considerations Cannot Reduce War's Hazard Alone," 827–828.

33 Michael Skerker, *The Moral Status of Combatants: A New Theory of Just War* (New York: Routledge, 2020).

34 Skerker, *The Moral Status of Combatants.*

35 See James M. Dubik, *Just War Reconsidered: Strategy, Ethics and Theory* (Lexington: University Press of Kentucky, 2016) especially Chapters 2–4.

8 Military Necessity and Realism

Comparing Permission and Limitation in Christian, Islamic, and Hindu Thought

Valerie Morkevičius

The idea of necessity is sometimes used to distinguish between realism and the just war tradition. Michael Walzer, for example, contrasts just war thinking with realism, which treats war as a "realm of necessity and duress" in which appeals to norms and values can never be more than "idle chatter."[1] Yet it would be a mistake to assume that realists are entirely uninterested in moral questions.[2] While the tale of Melos' destruction at the hands of the Athenians in the Peloponnesian Wars is often used as the example *par excellence* of realism at work, Thucydides himself seems to think the destruction was *not* necessary, and that this act of brutality reveals Athens' fall from moral grace into a doomed cycle of hubris, greed, and strategic overreach. Necessity, then, even for realists can serve both as permission and limitation.

Concern with necessity and duress also underlies just war thinking. Thus, for the Protestant theologian Paul Tillich, just war thinking demands finding "a way between a pacifism which overlooks or denies the necessity of power… and a militarism which believes in the possibility of achieving the unity of mankind through the conquest of the world."[3] Just war thinking within the historical traditions is by no means utopian or idealistic. Like realism, it is attuned to practical realities and underpinned by assumptions about the limitations and fallibility of human beings.

Indeed, as the pacifist theologian Stanley Hauerwas has pointed out, just war thinking invokes necessity in two ways: not only as a restraint (as when arguing that a particular war is unnecessary) but also as a permission, or even a duty (as when "the failure to go to war cannot but result in injustice").[4] This chapter aims to disambiguate these two profoundly different ways in which necessity functions as a principle. As Patterson has reminded us in his introductory chapter, while necessity restricts the scope of the just cause principle and underlies the principle of last resort, necessity is also used to explain why the use of force is ever justified in human affairs and to loosen some of the *in bello* principles that strategists and warfighters might otherwise find too idealistic. To this end, this chapter explores each of these faces of necessity through the lens of the historical just war traditions as they evolved within Christianity, Islam, and Hinduism.

DOI: 10.4324/9781003390398-8

Necessity and the Need for Force

In the Christian just war tradition, the idea of force as necessary in foreign relations arises from a broader conversation about the use of force and ultimately the need for political authority domestically. In Augustine's heavenly city, princes and judges would be unnecessary – with all hearts turned toward God, there would be no conflicts of interest needing to be resolved and certainly no evil-doers needing to be constrained. The earthly city, however, is populated by ordinary, fallen human beings who struggle to get along. And so earthly government is needed to keep order. The authorities in the earthly city set about organizing "the compromise between human wills in respect of the provisions relevant to the mortal nature of man," and "sanctioning with suitable vigor laws that enjoy just behavior and prohibit its opposite."[5]

Creating earthly order, in Augustine's view, requires coercion and even the use of force.[6] Augustine argues that:

> the duty of anyone who would be blameless includes not only doing no harm to anyone but also restraining a man from sin or punishing his sin, so that either the man who is chastised may be corrected by his experience, or others may be deterred by his example.[7]

For Augustine, force is not inherently a sign of animosity: a virtuous man "loves even his enemies; and such is his love even for those who hate and disparage him, that he wishes them to be reformed so that he may have them as fellow citizens, not of the earthly city, but of the heavenly."[8]

In his view, the just prince, judge, or even executioner is motivated by love for those they punish. The wise judge does not aim to harm, but rather seeks to heal "the wounds of sin," even if a "certain benevolent harshness" is required.[9] By reframing judgment and corporeal punishment as acts of love, Augustine seeks to reassure his fellow Christians that accepting positions of political authority in the newly Christianized Roman Empire would not be contradictory to their Christian duty. Indeed, he argues that such service is fundamentally necessary:

> In view of this darkness that attends the life of human society, will our wise man take his seat on the judge's bench, or will he not have the heart to do so? Obviously, he will sit; for the claims of human society constrain him and draw him to this duty; and it is unthinkable to hum that he should shirk it.[10]

Human fallibility means for Augustine that inevitably even a good and wise judge will sometimes mistakenly torture, imprison, or even execute the innocent, while the guilty will sometimes escape without punishment. Nonetheless, Augustine reminds his readers,

> all these serious evils our philosopher does not reckon as sins; for the wise judge does not act in this way through a will to do harm, but because

ignorance is unavoidable – and yet the exigencies of human society make judgment also unavoidable.[11]

Soldiers, like judges, are instruments of order when they serve a just authority. In carrying out the law, they not only do their individual duty but also help protect the just order that political leaders are attempting to create:

> One who owes a duty of obedience to the giver of the command does not himself 'kill' – he is an instrument, a sword in its user's hand. For this reason, the commandment was not broken by those who have waged wars on the authority of God, or those who have imposed the death penalty on criminals when representing the authority of the State in accordance with the laws of the State, the justest and most reasonable source of power.[12]

Thus, just as the judge defends domestic order, soldiers defend the polity against outside threats. As Augustine counsels Boniface, the governor of the Roman province of Africa, a soldier should strive to "be a peacemaker, then, even by fighting, so that through [his] victory [he] might bring those whom [he defeats] to the advantages of peace."[13] Just as a virtuous judge is one whose actions are motivated by love, a soldier's virtue is determined on the basis of his intent. In this case, Augustine cautions that right intent requires fighting for the sake of peace. A soldier thus motivated will – like the judge on the domestic front – act out of necessity rather than personal interest: "Let necessity slay the warring foe, not your will."[14]

Later theologians shared Augustine's view that the nature of mankind necessitates the use of force for the sake of maintaining order both domestically and internationally. But while Augustine's argument justifying the use of force by earthly powers focused on the fallen nature of mankind, Aquinas – writing a millennium later – follows the ancient Greek philosophers in claiming that the state is necessary because of man's social nature. Humans desire to live in community, but cannot work together efficiently without a leader, just as "a group of men in a boat cannot pull together as one unit unless they are in some measure united."[15] The earthly polity that the prince's leadership organizes naturally needs protection. And so, as Aquinas puts it, it is thus lawful for the prince to use the sword both "in defense of the commonwealth against those who trouble it from within" and "to protect the commonwealth against enemies from without."[16]

Likewise, Vitoria asserted that human societies "cannot exist without some overseeing power or governing force."[17] Without a leader, the individuals within any polity would compete against each other, striving for their own advantage, while undermining the common good. Thus, secular power arises from a "utility and necessity so urgent that not even gods can resist it."[18] Legitimate authorities must uphold laws – both human and divine – in order to maintain peace within society. For Vitoria, this logically means that the polity has the power "to compel any who breach that peace and contain them in the bounds of civil duty."[19]

Similar attitudes about the necessity of force as a tool in political life can be found in the Islamic tradition as well. As the ninth-century philosopher Al Farabi asserted: "To prevent many [from behaving unjustly], there is need to inflict evils and punishments… When it is excessive, that is an injustice upon him personally; and when it falls short, that is an injustice upon the inhabitants of the city."[20] Several centuries later, ibn Taymiyya (who was more or less Aquinas' contemporary) likewise followed the Greeks in arguing that man is "civil by nature," and that any group by nature needs a leader to coordinate its actions.[21] Thus, the ruler is permitted to establish human laws that address earthly concerns not covered by Qur'anic law: "The ordaining of what is fitting and the proscription of the improper is completed only by means of legal penalties, for God curbs through ruling power (*sultan*) what He does not curb through the Qur'an."[22] Because of the vital importance of maintaining order within society, political authorities are permitted to use force: "God does not permit, in effect, to put to death certain creatures except in view of the public good. He said: 'Discord is more frightening than death.'"[23]

Likewise, the historian Ibn Khaldun, writing a generation after Ibn Taymiyya, argued that

> evil is the quality that is closest to man when he fails to improve his customs and when religion is not used as the model to improve him… evil qualities in man are injustice and mutual aggression. He who casts his eye upon the property of his brother will lay his hand upon it to take it, unless there is a restraining force to hold him back.[24]

Because political authorities are entrusted with maintaining order, they are obligated to restrain those who would undermine that order – by force, if necessary.

The idea that human polities are kept from chaos by the use of force is not unique to the Abrahamic traditions. Within the Hindu tradition, order itself is intimately connected with force and violence: "the existence of all fine and noble life, of higher morality, of all happiness, of all order, depends entirely on the basis of force."[25] This idea is encapsulated in the words of Lord Krishna, urging Arjuna to fight his unjust and corrupt kinsmen in the *Bhagavad Gita*: "These worlds would collapse if I did not perform action; I would create disorder in society, living beings would be destroyed."[26] The necessity of using force to uphold order translates into a special duty incumbent on the warrior class: the obligation to protect society. The *Code of Manu*, like other *dharmasutras*, makes it clear that this class is duty-bound to protect the people, both by enforcing domestic laws and guarding against outside incursions.[27] As in Christianity and Islam, force is understood to be a necessary tool of statecraft.

Necessity and Just Cause

While the idea of necessity provides the fundamental permission for using force in a general sense, necessity also helps to limit the ends to which such force can be invoked. Without any additional limits, the idea that a just cause *permits* fighting can easily slide into the belief that a just cause *requires* fighting – without any consideration of the consequences. Although the just war traditions accept the proposition that force may be useful in restraining or even preventing a variety of serious injustices, they differ from holy war traditions in that they are inherently consequentialist.[28] The just war thinker, therefore, considers not only whether or not there is a just reason for fighting but also more practical questions, such as whether the harms caused by using force in a particular case are proportionate to the good for which one would be fighting and even whether there is a reasonable chance that the use of force would succeed in its aim.

But the just war tradition does not rely on these additional consequentialist principles alone to restrain the rush to war. The way the idea of just cause is framed in the classical texts appeals to the concept of necessity to put limits on just cause itself. Augustine's treatment of just war is notably less formal than that of his successors, but he makes it clear "that necessity, not choice, should be the reason for killing an enemy."[29] Thus, Augustine underlines the need to use force in a passage justifying the wars of the Romans: "When they were subjected to unprovoked attacks by their enemies, they were *forced to resist* not by lust for glory in men's eyes but by the *necessity* to defend their life and liberty."[30] A just cause is one which is fundamentally forced upon us.

Gratian emphasizes the connection between just cause and necessity by asserting that any war is unjust if it is fought "by choice, not by necessity."[31] Aquinas further develops the principle of just cause, to assert that "those against whom war is to be waged must deserve to have war waged against them because of some wrongdoing."[32] While he acknowledges that ideally, a Christian should be prepared not to resist or defend himself, "it is sometimes necessary to act otherwise than this for the common good: even, indeed, for the good of those against whom one is fighting."[33] Just cause and necessity are so interwoven that when Vitoria treats the question of permissible reasons and causes of just war, he makes it clear that only wrongs that urgently necessitate a forcible response can justify fighting. After all, "since all the effects of war are cruel and horrible... it is not lawful to persecute those responsible for trivial offences by waging war upon them."[34] Calvin, in a similar vein, clarifies that "if arms are to be resorted against an enemy, that is, an armed robber, they ought not to seize a trivial occasion, or even to take it when presented, unless they are driven to it by extreme necessity."[35]

Martin Luther makes this connection between just cause and necessity explicit. He urges princes to "Make the broadest possible distinction between what you want to do and what you ought to do, between desire and necessity, between lust for war and willingness to fight."[36] In Luther's view, even a war motivated by a just cause fails to be a just war unless those resorting to war can in good conscience say, "'My neighbor compels and forces me to fight, though I would rather avoid it.' In that case, it can be called not only war, but lawful self-defense, for we must distinguish between wars that someone begins because that is what he wants to do and does before anyone else attacks him, and those wars that are provoked when an attack is made by someone else. The first kind can be called wars of desire; the second, wars of necessity."[37]

Within the Islamic tradition, the idea of necessity also serves to bracket more and less urgent sorts of just causes. On the one hand, the logic of the Islamic worldview implied that the part of the world not governed according to Islamic law was inherently disordered, and thus inherently both suffering itself under an unjust system and potentially threatening world order. And so, it was necessary to struggle against the injustice posed by *dar al-harb* (literally "the world of war," in contrast to the *dar al-Islam,* "the world of peace"), whether through peaceful means of persuasion and invitation, or through the use of force. The use of force, however, was always seen as a lesser form of *jihad* (or *struggle)* in comparison to patient forbearance.

As with the Christian idea of righting wrongs as a potential just cause, this way of basing the legitimate reasons for war on the perceived injustices committed by the other could potentially be a very permissive structure. However, the idea of necessity was used here as well to serve as a brake. Thus, many Islamic scholars within the Sunni tradition argued that defensive wars fought to protect Islamic lands and peoples were necessary, while offensive wars were only voluntary.[38] This translated, in the work of Ibn Taymiyya, for example, into the designation of defensive *jihad* as "a duty of individual obligation for all the believers, even if they are not personally attacked. It is considered like a duty of solidarity and cooperative help."[39] The voluntary nature of offensive wars was such that commentators warned against initiating them unless victory were likely. Thus, al-Shaybani recommends a conciliatory policy of treaties and diplomacy in situations where "the inhabitants of the territory of war are too strong for the Muslims to prevail against them."[40] The Shi'i tradition is more restrictive: because the rightful Imam is hidden, earthly rulers lack the legitimate authority to call for offensive *jihad*, although defense of the community remains obligatory.[41] In this view, political authorities are serving as caretakers, doing what is necessary to maintain the community until the true Imam is revealed again.

Although the Hindu tradition does not delineate just causes in a parallel fashion to Christianity and Islam, leaving the decision of when and why to go to war in the hand of secular authorities entrusted with benefitting the community, the idea of necessity nonetheless serves here as a restraint as well. The underlying assumption is that "peace is the very basis for the existence and continuation of any political system," and that peace is the best means for assuring progress.[42] Thus, in the *Mahabharata*, the narrator asks "what is more frivolous than going to war? … Who cursed by his fate would choose war?"[43] War, then, is not something to be chosen – but rather an act of necessity.

Necessity and Temporality

One can imagine that the grounds for a just cause may persist for a very long time. Most wars do not occur suddenly, but instead arise out of long-stewing discontents and conflicts. So how do we know that *now* is the time to use force – *now*, as opposed to sometime in the future? In contemporary terms, this question is answered by the last resort criterion. Here too, we find the idea of necessity is woven through, just as it is tied up with the very question of what constitutes a just cause. In a sense, the need for a principle like last resort is implied by the way in which necessity serves as a limit on just cause. This is the argument Dietrich Bonhoeffer made in justifying his decision to participate in a plot to kill Adolph Hitler, despite his personal commitment to Christian pacifism.

> The destruction of the life of another may be undertaken only on the basis of an unconditional necessity; when this necessity is present, then the killing must be performed, no matter how numerous or how good the reasons which weigh against it. But the taking of the life of another must never be merely one possibility among other possibilities, even though it may be an extremely well-founded possibility. If there is even the slightest responsible possibility of allowing others to remain alive, then the destruction of their lives would be arbitrary killing, murder.[44]

If a particular use of force is truly necessary, then it becomes a duty. But for Bonhoeffer, the use of force and the killing it entails can never be treated as simply another tool of politics. Because of the gravity of the sin involved, it is a decision that can never be chosen *if there is any other way.*

While most just war thinkers do not share Bonhoeffer's belief that *all* killing is inherently sinful, the logic of last resort within the Christian just war tradition functions in essentially the same way. Unlike legitimate authority, just cause, and right intent, last resort is a relatively recent addition to just war thinking.[45] While James Turner Johnson argues that last resort's contemporary popularity is tied to our contemporary "uneasiness with modern war," its origins can be found in the works of the first Protestant thinkers.[46] Luther, for example, requires that one must "wait until the situation compels you to fight."[47] Similarly, Calvin claims that a just war for Christians requires going beyond the Ciceronian requirement of fighting for the sake of peace, arguing that Christian duty requires that "certainly we ought to make every other attempt before we have recourse to the decision of arms."[48]

Last resort is a very prominent part of the Hindu just war tradition. In the *Mahabharata,* several sages advise the heroes that "peace between the good should be sought at least once."[49] Indeed, nonviolent means of conflict resolution are held up as preferable to violent ones: "Victory gained by means of conciliation is said to be the best. When it is won by sowing dissension in the enemy ranks, it is intermediate, whereas victory won in a war is the worst kind."[50] Similarly, in the *Ramayana*, Ravana's brother counsels that "the learned have prescribed as appropriate the use of force only on those occasions where one's object cannot be achieved by means of the other three stratagems."[51] The Code of Manu reiterates the same principle: conciliatory strategies should be used first, with force as a tool to be used only if "they still do not submit."[52] In emphasizing the value of seeking alternate ways of resolving conflicts, these texts reinforce the idea that force should be used only if necessary – that is to say, only as a last resort.

Jus in Bello Necessity

The *jus in bello* rules for fighting justly emerged later in the just war tradition than the *jus ad bellum* ones, so the earliest canonical texts do not treat such issues in systematic detail. Yet even among those sparse references, the principle of necessity can be found. Taking up the problem of fighting on holy days, Aquinas draws a parallel between the ruler and a doctor – who must of course care for sick people whenever they need him. With this analogy in mind, Aquinas declares that it must be "lawful to wage just war on holy days, provided only that it is necessary to do so."[53] Preserving the "health of the commonwealth – to prevent the slaughter of many and innumerable other ills both temporal and spiritual" makes defending it an act of necessity.[54] Similarly, the early fourteenth-century canonist Johannes Andreae used the principle of necessity to justify forming alliances with non-Christians:

> I consider that pacific infidels, not having active or passive war with us, can in case of imminent necessity licitly be called upon, otherwise not, that is (not) if war is successful or if our war is unjust, or if it is just but necessity is not urgent.[55]

Even in these very early explorations of the principles of how to fight justly, the idea of necessity is already playing a role, primarily in its permissive aspect.

The principle of necessity is deeply implicated with the development of more detailed *jus in bello* principles at the beginning of the modern era. Vitoria uses the concept of necessity in regard to the principle of discrimination. The requirement to distinguish between civilian and military targets (and to target only the latter) comes up twice in Vitoria's opus on the laws of war. On the one hand, he considers the question of whether it is lawful to sack a city. He responds that "this is not of itself unlawful, if it is necessary to the conduct of war."[56] However, he tempers this permission with a reference to necessity. Setting fire to or destroying a Christian city could be permissible, but only "in the most pressing necessity and with the gravest of causes."[57] Given the implication in the surrounding text that such violence is sadly customary, necessity in this context appears to be acting as a restriction.

More famously, Vitoria elucidates the principle of double effect as a loophole to a strict interpretation of civilian immunity. In this case, however, necessity is functioning again as a permission, rather than a restriction. While Vitoria asserts that "it is never lawful in itself intentionally to kill innocent persons," he counters that "it is occasionally lawful to kill the innocent not by mistake, but with full knowledge of what one is doing, if this is an accidental affect."[58] By "accidental," Vitoria clearly understands something akin to the contemporary idea of military necessity:

> For example, during the justified storming of a fortress or city, where one knows there are many innocent people, but where it is impossible to fire artillery and other projectiles or set fire to buildings without crushing or burning the innocent along with the combatants. This is proven, since it would **otherwise** be impossible to wage war against the guilty, thereby preventing the just side from fighting.[59]

Vitoria emphasizes his point about necessity by reminding the reader that the innocent cannot be killed "even accidentally and unintentionally, except when it advances a just war *which cannot be won in another way*."[60]

Grotius also begins by asserting a broad principle of civilian immunity. Although civilians may be killed in the course of a legitimate military action, "yet humanity will require that the greatest precaution should be used against involving the innocent in danger, except in cases of extreme urgency and utility."[61] Ultimately, his argument, like those of his predecessors, hinges on military necessity: "No one can be justly killed by design, except by way of legal punishment, or to defend our lives, and preserve our property, when it cannot be effected without his destruction."[62]

Within the Islamic tradition, necessity serves a very similar function, enabling something like the principle of double effect to emerge. Al-Shaybani develops a virtually identical argument to Vitoria's (although writing some seven hundred years earlier). Although innocents should not in general be harmed, their indirect deaths as the result of targeting military objects – including city walls – could be justified. Al-Shaybani thus explains that "if the Muslims stopped attacking the inhabitants of the territory of war for any of the reasons you have stated, they would be unable to go to war at all."[63] Similarly, al-Shafi'i denies that "a habitation that is otherwise permissible to attack becomes forbidden due to the presence of a Muslim, whose blood is inviolable, within it," although he cautions that it would be "reprehensible" to attack such a site as a mere "precaution."[64] Ideally, the attackers should leave the infidel city alone "until it is possible to fight them without their human shields."[65]

In a similar vein, the Shi'i scholar al-Muhaqqiq also asserts that civilians should be immune from attack, within the limits of necessity: "it is recommended to abstain from cutting trees belonging to the enemy, using incendiary projectiles and intercepting flowing water to deprive the enemy of it, except in the case of absolute necessity."[66] He argues that even if the enemy "covers his defensive front with women or children, one should avoid attacking him, insofar as it is possible, at least unless the battle has already begun."[67] Likewise, if the enemy makes use of Muslim prisoners as human shields, one should avoid killing them, even if attacking is a matter of necessity: indeed, al-Muhaqqiq asserts that if a Muslim soldier could have avoided killing a Muslim prisoner used as a human shield and still does kill him intentionally, he is morally guilty, and responsible for blood money or even retaliation.[68]

Within the Hindu just war tradition, by contrast, necessity sometimes serves as a permissive principle within the *in bello* context. On the one hand, the *Mahabharata* and *Ramayana* epics take an absolutist view, asserting that there are simply some people – women, Brahmin, children – that one may never kill.[69] Even when the other side is behaving unjustly, "evil cannot be met with evil."[70] Thus, in the *Ramayana*, Rama cautions his monkey general that the duty to protect a person who has come seeking safety is absolute, for "even at the cost of one's own life, a magnanimous person should save an enemy who has come for refuge."[71]

On the other hand, Kautilya – an early realist – permits the cutting off forts' drinking water and supply lines, but cautions against completely razing resistant strongholds on the grounds that if "a fort can be captured by fighting, fire shall not be used at all."[72] Meanwhile, the ordinary people should be encouraged to move away from the battlefield, and those who do should be granted favors and protection. For Kautilya, then, the concept of necessity seems to function in a similar way as it does in the Christian and Islamic traditions: a permissive principle that broadens the scope of permissible military activity beyond what a strict interpretation of the rules might seem to allow.

Conclusion

As this review of the various uses of necessity across three historical traditions reveals, the idea of necessity is quite close to the heart of just war thought. Such proximity should give rise to the recognition that the just war traditions are far closer to realism in their habits of thought than some contemporary thinkers in the Walzerian or revisionist schools might prefer. Like their realist counterparts, just war thinkers see politics in non-ideal terms: it is a realm inevitably marked by human fallibility, uncertainty, and strife. Thus, like realists, just war thinkers operating within these historical traditions do not seek to eliminate war, but rather to channel the use of force into directions that protect, rather than undermine, order and the common good.

This realization, however, should also serve as a caution. Awareness of the realities of power and politics can too easily slip into acceptance or even adulation. (Of course, the same temptation lies before realists, and paves the road to cynical Realpolitik.) Necessity is not a fixed point, but rather a guiding principle that can nudge us in multiple, sometimes contradictory directions. If we cede too much to necessity, we may find ourselves endorsing decisions that run counter to morality or counter to common sense or both. Yet treating necessity as an overly restrictive principle can lead to paralysis – and failing to act can be a moral failing in itself. A more self-conscious recognition of the role necessity plays in modulating our moral reasoning may help us avoid such excesses.

Notes

1 Michael Walzer, *Just and Unjust Wars* (New York: Basic Books, 2000), 4.
2 Valerie Morkevičius, "Power and Order: The Shared Logics of Realism and Just War Theory," *International Studies Quarterly* 59, no. 1 (2015), 11–22.
3 Paul Tillich, *Systemic Theology*, vol. 3 (Chicago, IL: University of Chicago Press, 1963), 387.
4 Stanley Hauerwas, "Should War Be Eliminated? A Thought Experiment (1984)", in John Berkman and Michael Cartwright, eds., *The Hauerwas Reader* (Durham, NC: Duke University Press, 2001), 417.
5 Augustine of Hippo, *City of God* (London: Penguin Books, 1984), XV.4, 599 and Augustine of Hippo, *Epistolae* 185.19 in Philip Wynn, *Augustine on War and Military Service* (Minneapolis, MN: Fortress Press, 2013), 182.

6 Hence Patterson's assertion that the just war tradition carries a primary presumption *for* justice, rather than against violence.

7 Augustine, *City of God*, XIX.16, 876.

8 Ibid., V.19, 212.

9 Augustine, "Letter 133, to Marcellinus" and "Letter 138, to Marcellinus," in Ernest L. Fortin, Roland Gunn, and Douglas Kries, eds., *Augustine Political Writings* (Indianapolis, IN: Hackett, 1994), 246 and 209.

10 Augustine, *City of God*, XIX.6, 860.

11 Ibid.

12 Ibid., I.21, 32.

13 Augustine, "Letter 189, to Boniface," in Fortin, Gunn, and Kries, p. 220.

14 Ibid.

15 Thomas Aquinas, *"De regimine principum,"* I.1, in R. W. Dyson, ed., *Aquinas: Political Writings* (Cambridge: Cambridge University Press, 2002), 5.

16 Thomas Aquinas, *Summa Theologiae*, IIaIIae 40, in Dyson, 240.

17 Francisco de Vitoria, "On Civil Power," 1.2, in Anthony Padgen and Jeremy Lawrance, eds., *Vitoria: Political Writings* (Cambridge: Cambridge University Press, 2003), 9.

18 Ibid., 10.

19 Ibid., 17.

20 Abu Nasr Al-Farabi, in Charles E. Butterworth, trans., *Alfarabi: The Political Writings; "Selected Aphorisms" and other texts* (Ithaca, NY: Cornell University Press, 2004), 41.

21 Ahmad ibn 'Abd Ibn Taymiyya *Public Duties in Islam: The Institution of the Hisba*, Muhata Holland, trans. (Leicester: Islamic Foundation, 1985), 20.

22 Ibn Taymiyya *Public Duties*, 60.

23 Ahmad ibn 'Abd al-Halim ibn Taymiya, *Le Traité de Droit Public D'Ibn Taimiya: Traduction annoté de la Siyasa sariya,* trans. Henri Laoust (Beirut: Institut Français de Damas, 1948), 128–129. My translation.

24 Ibn Khaldûn, trans. Franz Rosenthal, trans., *The Muqaddimah: An Introduction to History* (Princeton, NJ: Princeton University Press, 1967), 97.

25 Madanlal A. Buch, *The Principles of Hindu Ethics* (Baroda, India: "Arya Sudharak" Printing Press, 1921), 353.

26 Barbara Stoler Miller, trans., *The Bhagavad Gita: Krishna's Council in Time of War* (New York, NY: Bantam Books, 1986), III.24, 44.

27 Ankush R. Sawant, *Manu-Smriti and Republic of Plato: A Comparative and Critical Study* (Bombay, India: Himalaya Publishing House, 1996), 44–48.

28 Marc LiVecche's chapter is a good example of this observation.

29 Richard Shelly Hartigan, "Saint Augustine on War and Killing: The Problem of the Innocent," *Journal of the History of Ideas* 27, no. 2 (1996), 195–204, 201.

30 Augustine, *City of God.* IV.2. p. 137. Italics mine.

31 Gratian. *Decretum.* Part II, Causa 23, Question 2. In Gregory M. Reichberg, Henrik Syse, Endre Begby, eds., *The Ethics of War* (Wiley-Blackwell, 2006), 113.

32 Aquinas, *Summa Theologiae* IIaIIAe40, in R. W. Dyson, ed., *Aquinas: Political Writings*, 240.

33 Ibid., 241.

34 Vitoria, "On the Law of War," 1.3 in *Vitoria: Political Writings*, 303.

35 John Calvin, *Institutes of the Christian Religion*, vol. 2, John Allen, trans., Philadelphia Presbyterian Board of Publication and Sabbath School Work, 1902. XII, 646.

36 Martin Luther, "Whether Soldiers, Too, Can Be Saved," in J. M. Porter, ed., *Luther: Selected Political Writings* (Philadelphia, PA: Fortress Press, 1974), 113.

37 Ibid., 113.

38 Ahmad ibn 'Abd al-Halim Ibn Taymiyya *Le Traité de Droit Public d'ibn Taimiya: Traduction annoté de la Siyasa sariya*, Henri Laoust, trans. (Beirut: Institut Français de Damas, 1948), 134. See also Rudolph Peters, *Jihad in Classical and Modern Islam*

(Princeton, NJ: Markus Wiener Publishers, 1996), 53–54. Asma Afsaruddin points out that religious scholars outside the circle of political power only endorsed defensive *jihad*; those who wrote about offensive *jihad* tended to be closer to the seat of secular power. Asma Afsaruddin, *Striving in the Path of God: Jihad and Martyrdom in Islamic Thought* (Oxford: Oxford University Press, 2013).

39 Ibn Taymiyya *Le Traité de Droit Public d'ibn Taimiya*, 133. Translation mine. See also Peters, *Jihad in Classical and Modern Islam*, 53: "But if the enemy wants to attack the Muslims, then repelling him becomes a duty for all those under attack and for the others in order to help them."

40 Muhammad ibn Hasan al-Shaybani, *The Islamic Law of Nations*, Majid Khadduri, trans. (Baltimore, MD: Johns Hopkins University Press, 1966), V.603, 154.

41 See Valerie Morkevičius, "Shi'i Perspectives on Just War," in Howard M. Hensel, ed., *The Prism of Just War: Asian and Western Perspectives on the Legitimate Use of Military Force* (Ashgate Press, 2010), 145–168.

42 Bharati Mukherjee, *Kautilya's Concept of Diplomacy: A New Interpretation* (Calcutta, India: Minerva Associated Publications Pvt. Ltd., 1976), 30.

43 *The Mahabharata*, vol. 2, A. B. van Buitenen, trans. (Chicago, IL: University of Chicago Press, 1975), 5(50)26.1, p. 233.

44 Dietrich Bonhoeffer, *Ethics* (New York, NY: Touchstone, 1995).

45 James Turner Johnson, "Just War, as It Is and Was," *First Things*, January 2005.

46 Ibid.

47 Luther, "Whether Soldiers, Too, Can be Saved," 113.

48 Calvin, *Institutes of the Christian Religion*, vol 2. XII, 646.

49 *The Mahabharata*, vol. 2, 10.20, 71.

50 *Mahabharata, Book Six, Bhisma, Volume 1*, Alex Cherniak, trans. (New York: New York University Press, 2008), 3.80, 39.

51 *The Ramayana of Valmiki, Volume VI: Yuddhakanda*, Robert Goldman, Sally J. Sutherland Goldman, and Barend A. van Nooten, trans. (Princeton, NJ: Princeton University Press, 2009), VI.102.35, 35.

52 *The Law Code of Manu*, Patrick Olivelle, trans. (Oxford: Oxford University Press, 2004), 7.107–108, 114.

53 Aquinas, *Summa Theologiae* IIaIIae 40, art 4, responsio. In Dyson, 247.

54 Ibid.

55 Johannes Andreae ("Additiones Speculi: De Iudeis et Sarracenis") cited in Paulus Vladimiri, *Saevientibus* (1415), in Ludwik Ehrlich, ed. and trans., *Works of Paul Wladimiri*, vol. 1 (Warsaw: Instytut Wydawniczny Pax, 1968), 75.

56 Vitoria, "On the Law of War," 3.7.52, 323.

57 Ibid.

58 Ibid., 3.1.35–37, 314–315.

59 Ibid., 3.1.37, 315.

60 Ibid. Italics mine.

61 Hugo Grotius, *Rights of War and Peace*, A.C. Campbell, trans. (London: B. Boothroyd, 1814), III.11.

62 Ibid., III.11

63 Al-Shaybani, *The Islamic Law of Nations*, II.117, 102.

64 Al-Shafi'i, *al-Umm*, in Nesrine Badawi, "Sunni Islam, Part I: Classical Sources," in Gregory M. Riechberg, Henrik Syse, and Nicole M. Hartwell, eds., *Religion War and Ethics: A Sourcebook of Textual Tradition* (Cambridge: Cambridge University Press, 2014), 301–369, 337.

65 Ibid.

66 Ja'far ibn al-Hassan al-Hilli al-Muhaqqiq al-Awwal, *Droit Musulman: Recueil de lois concernant Les Musulmans Schyites*, A. Querry, trans. (Paris: L'Imprimérie Nationale, 1861), 326. I.9.2.2.40-41.

67 Al-Muhaqqiq, 326. I.9.2.2.42.
68 Ibid., 326. I.9.2.2.44.
69 Consider for example *The Mahabharata*, vol. 1, 1(9)143.1, 300 and I(19)149.10, 309 as examples.
70 *The Ramayana of Valmiki*, Volume VI, VI.102.35, 450.
71 Ibid., VI.12.15, p. 149.
72 Kautilya, *The Arthasastra*, L. N. Rangarajan, ed. and trans. (New York: Penguin Books, 1992), 13.4.50–53, 697–698 and 13.4.9–13, 694.

9   The Military Necessity of Ethics

Shannon E. French and Jonathan French Flint

People with the least understanding of the realities of modern warfare are often the first to excuse ethical transgressions on the grounds that they were unavoidable and emerged inevitably from the "fog of war." They are eager to appear either worldly and "in the know" or deeply sympathetic to the troops defending their best interests. In truth, they are neither. A significant number of unethical acts—including actual war crimes and atrocities—cannot in any way be defended as driven by military necessity. They do not help to win wars. Nor does suggesting that they do in fact help or support the troops.

There are two distinct points here that require separate justifications but strongly reinforce one another. First, there is the point that committing unethical acts in armed conflicts actively harms troops by causing psychological trauma and moral injury and undermining healthy transitions and readiness. If someone were to insist that these considerations are not enough to overcome the pressure of military necessity, they would then crash into the second point of the argument, which asserts (with the support of real-world examples) that military necessity demands no such "gloves off" approach. In contrast, abandoning restraints in modern warfare is a path to strategic failure. Military necessity is not at odds with ethics. Ethics are a military necessity.

How Ethical Breaches Harm Troops

Sgt. Sammy Davis, a US Medal of Honor recipient for his heroic actions in the Vietnam War, was a guest speaker at the US Naval Academy in the early 2000s. On one occasion, he spoke to the students (midshipmen—future Navy or Marine Corps officers) in the popular elective Ethics course, "The Code of the Warrior," telling them about his experiences and taking questions. Clearly hoping to undermine the course's themes and instructor, one midshipman, we'll call him Tom, took his shot with something close to this:

> All semester, our professor has been talking to us about the importance of preserving our humanity in war. But you've lived through the realities of combat. Isn't the truth that, as an officer, I should not waste time worrying about the humanity of my troops? My only job is to keep them alive.

DOI: 10.4324/9781003390398-9

The instructor describes what happened next:

> I held my breath. Sgt. Davis now had the power to completely undo every-thing I had tried to accomplish as an ethics instructor that semester—not to mention potentially shatter my own faith in the material I had been teach-ing and writing about for so long. I need not have worried. Sgt. Davis's response did more to encourage my midshipmen to take military ethics seriously than anything I had (or ever could have) done in the classroom, before or since.
>
> Sgt. Davis went right up to Tom, and shouted at him as only a sergeant can, "If that's what you believe, you do not deserve to be an officer, and you need to get out of my military right now!" After that opening blast, he went on with great passion to instruct Tom and all the other midshipmen present that, as officers, they must do everything in their power to safeguard the humanity—and not only the lives—of their troops. War is always an assault on the humanity of every individual caught up in its destructive path. That assault must be resisted as much as any physical threat. The men and women you lead into combat are your responsibility and ensuring that what you lead them to do does not strip them of their humanity is critical to discharging your fundamental duties as an officer.[1]

It turns out that Sgt. Davis's instincts, unsurprisingly, were sound. There is con-siderable support now among those who study the welfare of combat troops that there are few if any mental, emotional, or spiritual harms more damaging—both lastingly and sweepingly harmful—than moral injury. Moral injury is generally defined as a traumatic response to the violation of core values that cannot be rec-onciled or justified by circumstances. This violation may have been perpetrated by the sufferers or their leadership (or both).

One of the best-known scholars on this subject is psychiatrist Jonathan Shay, whose seminal work *Achilles in Vietnam: Combat Trauma and the Undoing of Character* opened many modern eyes to a phenomenon that is as old as war it-self. Shay asserts that there are certain kinds of "catastrophic war experiences that not only cause lifelong disabling psychiatric symptoms but can ruin good charac-ter."[2] These experiences are not only violent or shocking to the system but involve centrally the "betrayal of 'what's right.'"[3] The concept is studied across several disciplines, with similar conclusions. A recent integrative review published in the *Journal of Traumatic Stress* drew the following conclusion:

> Rooted in the self-perceived transgression of core personal convictions and values, which are often imbued with social or sacred importance, perpetra-tion, and betrayal-based moral injuries, can have a devastating impact on the emotions, relationships, health, and functioning of affected individuals.[4]

Strikingly, warfighters who have experienced moral injury due to their own actions that violated moral norms agree with Sgt. Davis about the stakes involved, including the recognition that some fates are worse than death. Consider this excerpt from a 2001 candid interview with then Senator Robert Kerrey, a former US Navy SEAL:

> As an inexperienced, 25-year-old lieutenant, Kerrey led a commando team on a raid of an isolated peasant hamlet called Thanh Phong in Vietnam's eastern Mekong Delta. While witnesses and official records give varying accounts of exactly what happened, one thing is certain: around midnight on February 26, 1969, Kerrey and his men killed at least 13 unarmed women and children. The operation was brutal; for months afterwards, Kerrey says, he feared going to sleep because of the terrible nightmares that haunted him.
>
> The restless nights are mostly behind him now, his dreams about Vietnam more reflective. One of those, which he says recurs frequently, is about an uncle who disappeared in action during World War II. "In my dream I am about to leave for Vietnam," Kerrey wrote in an e-mail message last December. "He warns me that the greatest danger of war is not losing your life but the taking of others' and that human savagery is a very slippery slope." [Kerrey] says he has spent the last three decades wondering if he could have done something different that night in Thanh Phong. "It's far more than guilt," he said… "It's the shame. You can never, can never get away from it. It darkens your day. I thought dying for your country was the worst thing that could happen to you, and I don't think it is. I think killing for your country can be a lot worse. Because that's the memory that haunts."[5]

There is an even more extreme response to moral injury that must be confronted. Senator Kerrey and others in similar circumstances who suffer remorse have not lost their humanity. If they had, they would have no regrets and no empathy for their victims. Some who experience moral injury choose, consciously or unconsciously, not to mourn the violation of their values but to reject those values themselves. This is a profoundly self-destructive path. Essentially, such individuals decide that they should never have attempted to follow any rules or norms in the first place and embrace a cynical or even nihilistic view of life and their role as a combatant. Such attitudes can spread like a virus through units, as new recruits are told that the rules exist only for show, written by lawyers far from the front lines. By rejecting boundaries that were drawn to protect their humanity, troops that buy into this jaded perspective cut themselves off from the ability to heal from their trauma and make healthy transitions back to civilian life.

Some of the more visible and tragic outcomes of moral injury include self-harm, substance abuse, and suicide. An extensive clinical study published in 2021 on the relationship between what the authors termed "PMIE," or "Potentially Morally Injurious Exposure" and suicide rates among serving military and veterans produced the following stark conclusions:

> Overall, findings of this study suggest that even after accounting for a host of factors including mental health symptoms, PMIE exposure due to

perpetration is a risk factor for men's suicide attempts during and after military service, and PMIE exposure due to betrayal is a risk factor for both women's and men's suicidal attempts during military service, but only women's suicidal attempts after service. Other important suicide risk factors include pre-military suicidal ideation and attempts, depression, PTSD symptoms, and MST. Of note, for men, while the relationship between PTSD and post-military suicide attempts weakened and was no longer significant (from peri- to post-military), the relationship between PMIE exposure by perpetration and post-military suicide attempt became stronger.[6]

As Shay further explains in "Casualties," the consequences of moral injury are in the end fully as destructive as a catastrophic physical injury:

> I want to dispute the habitual mind-body distinction that I myself implicitly made early in this essay by distinguishing physical from psychological injuries. This distinction is often useful, but at its root, incoherent. "The body keeps the score," as traumatologist Bessel van der Kolk has so resonantly said. The body codes moral injury as physical attack and reacts with the same massive mobilization.[7]

Philosophers Jessica Wolfendale and Matthew Talbert explain in *War Crimes: Causes, Excuses, and Blame* that except in rare cases (such as genuine psychopathy), men and women who participate in atrocious acts during their military service are not able to disassociate themselves—their moral selves—from the actions and responsibility for them:

> while perpetrators' dispositions, goals, beliefs, and values are affected by environmental factors, they are still attributable to perpetrators and are reflected in, and expressed through, their behavior. This is what gives perpetrator behavior its interpersonal moral significance for those affected by it, and what ultimately licenses victims' blaming responses.[8]

Although they may attempt to compartmentalize their actions or set aside their combat experiences as occurring within some kind of separate moral sphere, ultimately they cannot escape the reality that they were perpetrators of crimes against humanity.

Committing war crimes is an obvious path to moral injury. Those who lead others to the perpetration of such acts may, as Kaurin notes in her earlier chapter, motivate their followers by false appeals to military necessity, usually combined with abhorrent dehumanization of the enemy. The role of dehumanization in war is complex, and it may be that some degree of at least detachment or empathy dampening is required in order for any non-sociopath to perform their role as a combatant. The level of dehumanization that precedes war crimes, however, is significantly more extreme and usually relies on associating victims either with subhuman or superhuman (and again non-human) creatures (e.g., vermin or demons): "Systematic devaluation of the victim provides a measure of psychological justification for brutal treatment of the victim and has been the constant accompaniment of massacres, pogroms, and wars."[9] It also produces levels of aggression that are directly at odds with the fundamental military necessity of good order and discipline.

General Benoit Royal calls out the disastrous loss of discipline that accompanies such dehumanization in *The Ethical Challenges of the Soldier*:

> The soldier at war will always be liable to be overwhelmed by passion, a feeling of revenge, and the appeal of cruelty. In armies worthy of the name, it is right to require those who exercise command, at every level, to contain possible excesses of passion by their subordinates; for similar but more important reasons, it is essential that they prevent themselves using such excesses as a way of dramatically increasing their fervor in combat. …[T]he essence of the profession of arms [is]…the responsibility that the leader accepts for the use of force and the management of lethal risk.[10]

As World War II veteran and celebrated author J. Glenn Gray further vividly describes in his autobiographical book, *The Warriors: Reflections on Men in Battle*, the effect of this dehumanization of the enemy is incredibly far-reaching:

> The ugliness of a war against an enemy conceived to be subhuman can hardly be exaggerated. There is an unredeemed quality to battle experienced under these conditions, which blunts all senses and perceptions. Traditional appeals of war are corroded by the demands of a war of extermination, where conventional rules no longer apply. For all its inhumanity, war is a profoundly human institution (…). This image of the enemy as beast lessens even the satisfaction in destruction, for there is no proper regard for the worth of the object destroyed (…). The joys of comradeship, keenness of perception, and sensual delights [are] lessened (…). No aesthetic reconciliation with one's fate as a warrior [is] likely because no moral purgation [is] possible.[11]

Here again, the harms extend past the immediate experience of violations of the rules of war and can leave troops with long-term or permanent psychological wounds. This is no longer a controversial claim. It is even the case that, as military ethicist David Whetham notes, "governments are now held more accountable for the treatment of their soldiers in respect to basic human rights, both by society in general, and by their own legal systems."[12] Given that crossing ethical "red lines" wreaks such havoc on everyone involved, all that is left to consider is a so-called "dirty hands" argument that military success requires this extreme sacrifice of not only combatants and non-combatants, but all participants' mental and moral health. This is an empirical claim that demands an examination of whether such violations in fact support strategic ends.

How Ethical Violations Undermine Military Strategy

Carl von Clausewitz, arguably the most influential strategic theorist in the West, tells us that the physical and the moral are interconnected in warfare: "The art of war deals with living and with moral forces."[13] Accepting this contention means accepting that the moral is part of the choices people must make to conduct a war, a "Special Military Operation," a peacekeeping engagement, or any other use of force that falls short of a comfortable definition of war. At times, this may seem counterintuitive, especially to anyone whose concept of war comes from films like *Patton* that reify the worst excesses of an unrestrained commander and other media creations that condone the pardoning of murderers (those who have killed unjustly outside of the mandate given to them by the state as combatants) or torturers. Yet real-world conflicts consistently support it.

Clausewitz's clear lesson is that the ethical dimensions of war are inescapable. This is intrinsic to his conception of how wars are analyzed, how they are fought, and how they are won. To disregard it is to make the achievement of victory more difficult, to invite other players to the field, and to put leaders in peril of grim judgment once the dust has settled and the final accounting begins. It is often assumed, especially by non-specialists, that those prosecuting a war must discharge their dismal duty without paying attention to the niceties of ethical decision-making. This is the path to victory, it is thought—to pursue the end effectively and with efficiency. However, doing so brings with it detrimental consequences at the tactical, operational, strategic, and grand strategic levels of military decision-making.

To begin with the tactical level, the level at which the tip of the spear is brandished, these negative consequences can be seen in the effects on those sent to fight, their opponents, and those caught between the warring parties. Imagine your state is asked to intervene in a conflict that is small but bloody, with an enemy worthy of respect. Your state's will to expend resources for conflicts abroad is generally limited. The conflict in question takes place far away and is reported to you, but not experienced by you. There is no direct threat to your state, but the conflict imperils people with whom you are sympathetic and allied (if perhaps not officially obligated). You may feel a moral compunction to act, to resolve the situation by sending a limited number of resources (arms, troops, etc.) to help protect your allies. This impulse is laudable and to be praised. However, if charged with the mission of repulsing the enemy with limited forces and the best of intentions, it is easy to see trouble ahead. There will be a natural desire to save as many allies as possible, and shorn of restraint by exigency, there will be the temptation to use the maximum force available, without obeying the rules of war.

Firstly, consider the effect on your own troops. The maximum use of force, unconstrained by international norms, invites immediate considerations of those troops experiencing traumatic and post-traumatic stress disorder (PTSD) and moral injury. As previously explained, PTSD and moral injury are ethical concerns, but they are also tactical and operational issues. A psychologically injured soldier is an injured soldier. Those with visible injuries are treated, those without may not be. Anyone bearing these injuries may no longer be as reliable as expected and may even "go rogue." This was the defense used in the "Marine A"/Sgt. Blackman case in the United Kingdom,[14] who was accused of having killed a wounded Taliban fighter to reduce the charges against him from murder to manslaughter.

Sergeant Blackman had served several combat tours, including three in Iraq and two in Afghanistan. He had witnessed many horrors of war, and undoubtedly dishonorable behavior on all sides. As a result, he fell into the mindset of "anything goes," without understanding that the rules of war were put in place to protect him from sacrificing his own humanity. The actions of others are irrelevant. Perhaps your enemy tortures prisoners. This does not mean you can with moral impunity torture him. The reasons for you not to torture are tied to your values, not his. The issue is not what the enemy does or does not do, but what your own code demands.[15]

Whatever else Sgt. Blackman was, at the point where he took that life, he was not a benefit to his unit, the mission, or his state. Psychological stressors beyond the breaking point of the individual will render that individual less effective or indeed a hindrance to the mission. PTSD and moral injury are tactical issues in that they affect the reliability and availability of forces to do the business of fighting. Reducing the available fighting force from limited down to restricted is not a way to win. Readiness is a vital element of military necessity that is undermined by the unjust conduct of war.

Arriving with a force willing to fight the enemy at any cost presents further tactical challenges. Every strategist's favorite acronym in small wars, "WHAM" (which stands for "Winning Hearts And Minds"), becomes infinitely more difficult if a war is conducted without regard for rules and norms. Return to the imaginary conflict. If your state's troops act without honor and restraint, there will necessarily be innocent casualties. Your state then becomes part of the problem for the civilian population. Through civilian casualties and a perceived laissez-faire attitude to 'collateral damage' the trust in your forces will rapidly be diminished. This will be a propaganda bonanza for your enemies. Subsequent effects may include defection of the populace to the side of your enemy, loss of reliable intelligence, and the hatred even of those you were there to save.[16] Conversely, as proven by Colonel (later General) H.R. McMaster in Operation Iraqi Freedom, the softer approach, including engagement and provision of security, has an improving effect on all these matters.[17] Employing the WHAM policy, McMaster also saw losses among those he commanded decline and improvements in their readiness and availability. The advances he made in operational and tactical command were not, however, harbingers of innovation in higher command, so the gains did not last.

Not committing the "unforced error" of unrestrained use of maximum force denies the enemy the psychological space they need to thrive. It represents attacking one corner of the Claueswitzian trinity (which consists of forces of chance—the military, fear and enmity—the people, and logic and reason—the government) and destabilizing the enemy's center of gravity. This creates a significant tactical advantage by toppling the basis from which the other side of the conflict claims to fight. Exercising restraint makes it harder for your enemies to make their case, and easier for you to make yours.

Step back again into the imaginary conflict described above, no longer on the ground level, but on the next level up — the operational level. This is the level where decisions are made about which battles to fight and how they should be fought. Already on this level, there is a consequential effect that bubbles up from the tactical level. A force that is ready and uninjured is clearly better than an unready and injured one, providing more resources to apply to any upcoming battle. Acting with restraint may seem to cause a drop in overall efficacy in the short term, but the advantages will be reaped in the long term. If your troops have behaved well, in addition to suffering less from physical and moral injury, they will be able to work with a civilian population that is likely to view them with less distrust. This will make it easier to help the allies and gain support and even intelligence from the civilian population. The discipline and reliability of your state's forces will be more readily maintained with more troops available under less pressure. Without events such as unjustified killings that need explanation and prompt investigation and questioning, the overall morale of the contingent will hopefully be better than it otherwise would have been. Your enemy will be facing a disciplined force, less likely to overreact or make mistakes that can be capitalized on, while they may make errors of their own that produce opportunities. All this is the result of increasing the ethical standards from nil to at least a minimum level that includes and hopefully exceeds those codified in the Geneva Conventions.

Moving up now to the strategic level, this is all good news. This level is about enacting policy decisions, aimed at victory or mission accomplishment. Tactical and operational decisions kept within the boundary conditions of the norms of military restraint will deliver policy in a sustained and sustainable manner. Granted, this may take more time than an unrestrained, unethical approach, especially in a resource limited environment. But the only way to speed up the pace of such a conflict without adding additional resources would be to abandon restraint and risk strategic consequences and failure. In our scenario, it has already been suggested that your enemies, being capable and intelligent, will seek to drive to such choices—because they know no good can come of it for your state. Imagine the propaganda victory handed to an enemy should the worst befall a civilian village, either by mishap or intent. War is not conducted in a vacuum. Regional and international eyes are always there, and this is even more true with the democratization of information brought about by advances in mobile technology. At the strategic and grand strategic levels, forces must be aware that their actions can turn opinion with them or against them. The behavior—or misbehavior—of troops can call forth new allies or new enemies. Using the tip of the spear with elegance and thought (in modern terms, precision warfare with minimal collateral damage) will engender trust, but resorting to the broadsword or the blunderbuss will make allies turn away and foes (new or old) turn towards the enemy. At the grand strategic level (that of governments in organizations such as the United Nations), we risk the opprobrium of the international community and the squandering of any claim to moral authority we may have.

Moving away from thought experiments, history also supports these points on the convergence of ethics and strategy. A clear-eyed analysis of the Vietnam conflict will accept that the United States attracted old foes and handed the enemy horrific propaganda successes through the use of napalm and the indiscriminate Rolling Thunder bombing missions. Actions taken by US troops in Vietnam, including the well-publicized My Lai massacre, changed the perception of the US around the world. Similarly, it showed the effect of too few resources, attempting to do too much, with too little restraint and how this produced obstructive effects on the readiness and trustworthiness of US forces. A more recent example of this dismal spiral is detailed in the outcome of the Brereton Report into war crimes committed by the Australian Special Air Service.[18] The ongoing illegal and unethical Russian 'Special Military Operation' in Ukraine has also been disastrous on all these counts. Well-documented atrocities by Russian forces and Russian-paid mercenaries have managed to expand NATO, drive the Russian Army to mass conscription and the use of unreliable and outdated equipment while predominantly Western European philosophical states have supplied Ukraine forces with state-of-the-art tools of war and trained Ukrainian troops on their use. Russia has been isolated from international capital centers and inhibited by sanctions. All Russia's unrestrained behavior has done is to harden Ukrainian resolve and encourage their supporters. No matter the outcome of the war, Russia is unlikely to ever be able to claim any moral authority again, and will be treated with earned distrust by slightly wary friends and contempt by even more cautious competitors. As a test of the assertions above concerning the strategic cost of indefensible conduct of war, it could not be starker.

Conclusion

The nature of war has not changed. It is still bloody and dangerous and treacherous, but the modern character of war is one where the moral element will be decisive. A commitment to restraint, to observing ethical norms even in the most stressful circumstances, becomes a requirement at the tactical, operational, strategic, and grand strategy levels. Definitively, "taking the gloves off" is not a supportable or successful strategy in modern armed conflict, either from the perspective of maintaining troop readiness and mental health or that of achieving desired foreign policy goals. The psychological cost to troops, combined with the international reputational damage done to the perpetrators' state, cannot in any way be balanced by any (dubious) claims of military necessity. Beyond being manifestly unethical, "Kill them all and let God sort them out" is, beyond question, a losing strategy.

Notes

1 Shannon E. French, "Sergeant Davis's Stern Charge: The Obligation of Officers to Preserve the Humanity of Their Troops," *Journal of Military Ethics*, David Whetham, guest editor 8, no. 2 (2009): 116–117.
2 Jonathan Shay, M.D., Ph.D., *Achilles in Vietnam: Combat Trauma and the Undoing of Character* (New York: Simon and Schuster, 1994), xiii.
3 Shay, *Achilles in Vietnam*, xiii.
4 Brandon J. Griffin, Natalie Purcell, Kristine Burkman, Brett T. Litz, Craig J. Bryan, Martha Schmitz, Claudia Villierme, Jessica Walsh, and Shira Maguen, "Moral Injury: An Integrative Review," *Journal of Traumatic Stress* 32 (June 2019): 357–358.
5 Gregory L. Vistica, "One Awful Night in Thanh Phong," *New York Times Magazine*, April 25, 2001.
6 Shira Maguen, Brandon J. Griffin, Dawne Vogt, Claire A. Hoffmire, John R. Blosnich, Paul A. Bernhard, Fatema Z. Akhtar, Yasmin S. Cypel, and Aaron I. Schneiderman,

"Moral Injury and Peri- and Post-military Suicide Attempts among Post-9/11 Veterans," *Psychological Medicine* 53, no. 7 (May 2023): 3200–3209, https://doi.org/10.1017/S0033291721005274.

7   Jonathan Shay, "Casualties," *Daedalus: The Journal of the American Academy of Arts & Sciences* 140, no. 3 (Summer 2011): 186.

8   Jessica Wolfendale and Matthew Talbert, *War Crimes: Causes, Excuses, and Blame* (New York: Oxford University Press, 2018), 109.

9   Stanley Milgram, *Obedience to Authority: An Experimental View* (New York: Harper Perennial Modern Thought, 2009 edition), 9. For more current (and excellent) analysis of obedience in the military context, please see Pauline Shanks Kaurin, *On Obedience: Contrasting Philosophies for the Military, Citizenry, and Community* (Annapolis, MD: Naval Institute Press, 2020); and Nikki Coleman, *Obedience in the Military* (Routledge, 2020).

10   General Benoit Royal, *The Ethical Challenges of the Soldier: The French Experience* (Paris: Economica, 2012), 63–64.

11   J. Glenn Gray, *The Warriors: Reflections on Men in Battle* (New York: Harper and Row, 1970), 152–153.

12   David Whetham, "What Senior Leaders in Defence Should Know about Ethics and the Role That They Play in Creating the Right Command Climate," *The International Journal of Ethical Leadership* 8, no. 1 (2021): 73–93.

13   Carl von Clausewitz, *On War* (Princeton, NJ: Princeton University Press, 1989), 86.

14   See Martin L. Cook, "Military Ethics and Character Development," in G. Lucas, ed., *Routledge Handbook of Military Ethics* (New York: Routledge, 2015), 123–132.

15   Shannon E. French, "Revelation and the Rules of Engagement: Sultan Saladin and the Warriors of Islam," in *The Code of the Warrior: Exploring Warrior Values Past and Present,* 2nd ed. (Lanham, MD: Rowman & Littlefield, 2017), 235–252.

16   Matthew Alexander, "McCain Backs Torture as Recruiting Tool for Al Qaida; Policy Led to the Deaths of U.S. Soldiers in Iraq," *Huffington Post*, August 31, 2009.

17   T. Harford, "Lessons from War's Factory Floor," *Financial Times*, 23 May 2011.

18   The Inspector-General of the Australian Defence Force Afghanistan Inquiry, available at https://www.defence.gov.au/about/reviews-inquiries/afghanistan-inquiry (accessed 21/14/2023).

10 "Operation Wrath of God"

Illegal but Necessary

Amos N. Guiora[1]

Introduction

On September 5, 1972, eleven Israeli athletes were murdered by the Palestinian terrorist organization, Black September, at the Munich Olympics.[2] This event is widely understood to mark the day contemporary terrorism was born. Similarly, modern-day counterterrorism quickly followed when Israeli Prime Minister Golda Meir gave the order to kill those involved in the planning and execution of the attack. That decision will be the focus of this chapter.

The name 'Black September' references the events of September 1970 when King Hussein of Jordan forcefully attacked Palestinians living in Jordan, resulting in significant casualties, and subsequent claims of torture by Jordanian forces. As a result of Hussein's actions, many Palestinians left Jordan, making their way to refugee camps in southern Lebanon.[3]

The 1972 Munich Olympics were supposed to mark the "new" Germany and erase the memories of the heavily criticized 1936 Berlin Olympics where Hitler was provided an opportunity by the International Olympic Committee to present a positive and sanitized image of pre-World War II Nazi Germany. To show the world a "different" Germany in 1972, security at the Olympic Village was lax, a matter of tragic significance when examining the events of September 5. The purpose of introducing this information is that the irony of Israeli athletes being murdered on German soil is integral to the decisions made thereafter.

The question of whether "Operation Wrath of God" met the test of military necessity is of particular relevance to this analysis. The traditionally understood definition of military necessity focuses on whether the use of force is justified in responding to or protecting from an attack. The second part of the definition requires determining whether the use of force—if it meets the test of military necessity—is proportional to the threat and used discriminately. This analysis is intended to articulate the limits of power while recognizing that self-defense is legitimate. In examining "Operation Wrath of God," we must ask whether Prime Minister Meir's decision to kill the Black September operatives met the test of military necessity, informed by proportionality and discrimination, under the umbrella of legitimate self-defense, or, rather, was predicated on retribution and deterrence.

DOI: 10.4324/9781003390398-10

"Operation Wrath of God" poses the dilemma requiring us to differentiate between punishment and retribution. In deciding to kill those responsible for the murder of 11 Israeli athletes, Prime Minister Meir ordered the Mossad to kill—not capture—the Black September operatives. Had she ordered their detention leading to a trial in Israel (or in Germany) that decision would have resulted in the punishment of those responsible. In choosing to order the Mossad to kill them, Prime Minister Meir engaged in retributive conduct. There is no indication of available material that consideration was given to arresting those responsible for Munich and bring them to trial in Israel as was done with Adolf Eichmann.[4]

Background

The decision to attack the Israeli athletes reflected Palestinian anger with respect to the events of September 1970, and the 1948 Arab-Israeli War, as well as the 1967 Six-Day War involving Israel, Syria, Jordan, and Egypt.[5] In the aftermath of both wars, Palestinian refugees became a central issue in the politics and conflicts in the Middle East.

The Palestinians did not participate in the 1967 War; however, there was a direct consequence from their perspective in that rather than living under Jordanian rule (West Bank) and Egyptian rule (Gaza Strip) as they had from 1948–1967, they now lived under Israeli occupation, albeit Israel did not annex the West Bank. Rather, the Israel Defense Forces (IDF) assumed responsibility—in accordance with the international law of occupation—for Palestinians living both in the West Bank and Gaza Strip.[6]

To draw attention to their plight and condition, Black September leadership determined that the Munich Olympics presented a unique opportunity to put the Palestinian cause "front and center" on the global stage. They were absolutely correct about that assumption.

The crisis began with an early morning assault on the living quarters of the Israeli athletes resulting in two Israeli athlete deaths.[7] The ensuing demands and threats were clear: failure to guarantee the release of Palestinian prisoners by Israel would result in the killing of the remaining Israeli athletes and coaches being held hostage in their living quarters.

As the world watched via live TV from ABC News, German officials negotiated with the Black September terrorists who were demanding the release of Palestinian terrorists and prisoners held by Israel. The official Israeli policy at the time was to not negotiate with terrorists.[8]

While the deadline given by Black September moved throughout the day, Israeli Prime Minister Meir was adamant in her resolve to follow that policy, albeit the lack of direct interaction between German and Israeli officials rendered the resolve somewhat moot. To this day, this is a source of discussion and speculation. The head of Mossad, Zvi Zamir, traveled to Germany during the day to offer his advice but his efforts were rebuffed by German officials despite his experience and expertise.[9]

Late in the day, German officials agreed to transport the athletes and terrorists via helicopter to a local airport from which they would fly to an Arab country to effectuate an exchange thereafter.[10] The plan was a German ruse: at the airport, German snipers opened fire on the terrorists who in turn killed all the Israelis.[11] From an operational perspective, the mission was a resounding failure, reflecting poor planning, ill-equipped forces, and problematic command and leadership. The German officials refused to allow the participation of Israeli forces, eschewing advice and suggestions from senior Israeli officials who were on-site at the airport.[12]

The International Olympic Committee made the decision, after a memorial service and one-day abeyance, to continue the Games.[13]

Several Black September terrorists were killed during the German operation and the remainder were captured but were eventually released two months after the massacre when West Germany acceded to terrorist demands in the aftermath of the hijacking of a Lufthansa Boeing 727. At no point was Israel consulted regarding the situation.[14]

While other attacks exacted a greater price in terms of casualties, what makes Munich-"Operation Wrath of God" particularly compelling and important are the following:

- It was, literally, the first terrorist attack a world-wide audience could watch unfolding "live," including seeing one of the terrorists standing on the balcony;
- The comings and goings of German officials interacting with the terrorists;
- The bitter, historical irony of the location of the attack and its unique circumstance;
- ABC Sports commentator Jim McKay's words, which were heard by hundreds of millions;
- The clear determination by Israel to extract a steep price;
- The means by which Black September operatives were killed after their release; and
- The unapologetic motivation for the Israeli Operation.

"Operation Wrath of God"

When broadcaster Jim McKay, informed the world, "they are all gone," speaking of the lives of the Israeli athletes and coaches who were taken hostage and murdered, the operative question became what Israel would do in response.[15] Israel answered by implementing a governmental decision reflecting a combination of revenge and deterrence, which has become known as "Operation Wrath of God" ("Operation Bayonet" in Israel).

The operation was conducted by the Mossad, the Israeli national intelligence service that operates outside of Israel's borders—distinct from the SHABAK that operates within Israel, the West Bank, and the Gaza Strip. An operation of this magnitude required teams working in Europe, Cyprus, and the Middle East both to track targets and to kill them.[16]

There was no discussion of capturing, detaining, or interrogating the identified Black September operatives. The operational goal was clear and concise, devoid of nuance: kill those deemed responsible for the murder of the 11 Israelis. The killings were conducted in a variety of ways, including shooting, bombs placed in telephones and refrigerators, and under an individual's bed; the methods were intended both to kill the intended target but also to send a clear message to others. The operation was halted after the mistaken-identity killing of an innocent Moroccan waiter in Lillehammer, Norway.[17]

Notwithstanding this significant operational error, from an Israeli counterterrorism perspective the undertaking was both justified and effective. Despite the effectiveness, significant criticisms reflected the common consensus in international law that *revenge-based* operations are not tolerated. Vengeance is distinct from *self-defense*, a right granted to nation states in Article 51 of the United Nations Charter. While the UN Charter defines self-defense in the context of state-to-state conflict, it has now been applied by nation-states engaged in operational counterterrorism against terrorist organizations (involving non-state actors).[18]

Prime Minister Meir gave short shrift to questions of morality and self-imposed limits of state power. There is no suggestion that this was a relevant concern from her perspective; rather, her focus was *exacting* (word used deliberately) a price for the attack. By all accounts, this was made clear to the Mossad leadership and operatives.[19] In doing so, whether intended or not, Meir was framing the decision in the context of *military necessity*. Regardless of her specific articulation, *Meir determined that killing those responsible was necessary both to punish and to deter those planning future acts of terrorism.* The intended message being that the long arm of Israel will find you, regardless of your location, and will ensure your demise.

Message-Sending Motives

Nonetheless, *military necessity*—as applied by Israel in the aftermath of the Olympics—must be understood from the perspective of *message sending*. After all, the killing of the athletes did not pose either a practical or existential threat to Israel, either immediately after their death or in the long term. The operation, then, must not be viewed as a response to an act of war that threatened the nation-state; that traditional understanding and application of military necessity is not applicable to Prime Minister Meir's decision. Thus, the military necessity relevant to "Operation Wrath of God" must be framed different from in context of state-to-state conflict. While terrorism/counterterrorism is predicated on state/non-state conflict, "Operation Wrath of God" poses a distinct question regarding military necessity.

The question, from Prime Minister Meir's perspective, extended beyond the physical engagement with those responsible. While the operation was logistically and operationally complex, its importance extended well beyond the killing of specific individuals, notwithstanding their guilt and involvement. That "extension" is a direct reflection—from a public or moral perspective—of the significance of the attack (event, location, history, result), and the need for sending a message to terrorist organizations. In that sense, the military necessity needs to be understood from two distinct perspectives: (1) the "power" of the attack; and (2) the practical import of its undeniable success. Another way of thinking about this is that "Operation Wrath of God" must be understood from an intelligence/military perspective and, simultaneously, in terms of public messaging to disparate audiences.

After all, 11 Israeli athletes were murdered on a stage intended to celebrate both the "new" Germany and the beauty and elegance of sport and unity of the international community. This was not a "typical" terrorist attack; it was something larger, more dramatic, and significant. Without minimizing the loss of innocent life, from a quantitative perspective, 11 deaths were not "off the charts." Rather, what made it so powerful, perhaps searing, was the confluence of three different factors. First, as noted above, the day's events were broadcast internationally on live television, including Jim McKay's dramatic announcement: "They are all gone." Second, the momentous historical symbolism of another anti-Jewish Munich Olympics, combined with a third element, the stunningly incompetent rescue mission and its consequences, including Germany's callow surrender to terrorist demands.

That is reflected in the way the operation was conducted. In assessing whether the operation met standards of morality and military necessity, we need to examine the question from two distinct angles: Prime Minister Meir "in the moment," and retrospectively with the benefit of time and distance. To understand the Israeli public mood of 1972 requires recognizing that the trauma of the Holocaust was very much front and center.

That was magnified by the fact Israelis were murdered on German soil, in part—if not large part—because of German incompetence. The phrase, "out of the ashes of the Holocaust" is used to describe Israel's birth and (perhaps) raison d'etre. From this perspective, Prime Minister Meir's decision is understandable. However, the compelling circumstances of the attack do not inherently suggest the response meets either the test of morality and/or military necessity. The analysis of both, regardless of the death count and symbolism, needs to be analyzed independently. Whether Prime Minister Meir did that, herself, is an open question.

Assessing Morality

From all available information, Prime Minister Meir and the Mossad did not weigh questions of morality, but rather were focused on the need to respond with an iron fist. While that is understandable for the reasons suggested above, it must not blind us to the fact that historical and operational considerations of whether a state act is moral or immoral should not be dismissed out of hand. Regardless of whether one supports or condemns terrorism, a government's response is traditionally understood to be examined through a different lens. After all, the state has means, capability, and resources that far exceed that of non-state actors.

That, however, does not grant the state the right to operationalize its full arsenal when confronting non-state actors, whether proactively or reactively, in immediate response to an attack. "Operation Wrath of God" represents a different model, where state action was neither proactive nor reactive (in the immediate response model) but reflected revenge well after the event ended. Notwithstanding the then-head of the Mossad, Zvi Zamir, there was no intelligence information suggesting an imminent attack by Black September.

When discussing morality, there are three important considerations. The first is how to define moral action in a given context. The second is how to implement morally satisfying policies. The third is accepting the consequences of self-imposed standards of morality.

We begin with the definition of morality. For the sake of this understanding, morality is defined as ethical decision-making, unrelated to law, whether written or otherwise. The notion of ethical decision-making in operational counterterrorism arguably is suggestive of an oxymoron but, from the perspective of commanders, the failure to have clear codes of conduct has two important consequences: it gives license to one's soldiers to act immorally or a "license to kill" and enhances the vulnerability of the commanders' soldiers through "unintended consequences."[20]

While operational counterterrorism is predicated on state power directed at non-state actors, the engagement must (in accordance with the law of war and international law) be limited to the individual(s) who pose the imminent threat to the state, i.e., the principles of proportionality and discrimination. Otherwise, without these, the paradigm would be akin to the Wild West. Implementation depends on the commander's articulation to soldiers regarding the appropriate level of force, in accordance with the threat posed.

Self-imposed restraint—what former President (Chief Justice) of the Israel Supreme Court Aharon Barak called "fighting terrorism with one arm tied behind your back"—is the essence of morality in operational counterterrorism. However, while applauded in some circles, the model is not cost-free for there is the possibility that such an approach endangers one's own soldiers. Restraint, while intellectually appealing, can have negative practical consequences.

Whether Prime Minister Meir considered this dilemma is an open question. The sense from the available material is that she did not; rather, her focus was how to respond from an operational perspective most effectively to an event of this magnitude. In addition to the operation's success from an intelligence/military perspective, notwithstanding the tragic error in Norway, the importance of punishment and message-sending were also of paramount importance.

Measuring Effectiveness

Operational military necessity (*jus in bello*) is closely tied to the *jus ad bellum* criteria of likelihood of success, because tactical and operational level successes are directly tied to strategic decision-making and the goal of peace. Whether Wrath of God was a success depends on which lens the question is analyzed: whether based upon short-term or long-term criteria. From a societal "morale" perspective, by sending a clear message, the killings were arguably effective, albeit with muted criticism from some of the families of the Israeli athletes. From the perspective of a terrorist organization, the operation reflected significant capability in identifying, tracking, and killing the targets. For a terrorist organization that is a powerful reminder of their vulnerability and the power and ability of the state. In the context of operational counterterrorism that is important both practically and intangibly; the former because it reflects weakness, the latter because it can cause doubts within the organization.

An operation of this nature requires resources, commitment, and skill at the very highest level of counterterrorism. While the mistaken identity in Norway was an egregious error that rightfully led to cancellation of the operation, that terrible operational error should not mask the fact that the overall operation was a stunning success. When examining the question of morality and military necessity, that fact must not be forgotten or casually dismissed. The importance of demonstrating operational capability, thousands of miles from Israel, was a clear signal to terrorists that Israeli counterterrorism went well beyond its borders and natural environs. That is of major significance when examining the tension between effectiveness and military necessity; the former does not guarantee the latter. The two points need to be separately analyzed, rather than from the same perspective as if being run through the proverbial blender together. The positive attribute to one factor does not necessarily carry over to the second attribute. That is very much the case here.

From an effectiveness perspective (narrowly defined), Israel achieved its immediate purpose. One of the most impressive aspects of the operation was the extraordinary precision with which it was conducted, albeit resulting in one instance of collateral damage. While the killing of the Moroccan waiter in Norway was a major tragedy, the fact that he was the only innocent individual killed in an extremely complicated operation is noteworthy. Whether Wrath of God had a long-term impact on potential terrorists and future acts of terrorism is unknown.[21] What is clear is that the rationale for operational counterterrorism was expanded by Israel from a reactive self-defense model to a broader paradigm, crossing into a gray zone that was not in accordance with a narrow interpretation of international law or self-defense.

Quite the opposite. In undertaking this operation in the manner it did, Israel ignored—deliberately, as is clear—standards that are generally applied to actions of a state power to non-state actors. While the circumstances arguably, from the perspective of Prime Minister Meir, justified this expansion (unilateral and brazen), it ran counter to a more somber, cautious use of power. It is not an exaggeration to suggest that Israel knowingly ignored those rules and limits. Again, this demonstrates tensions between effective action, morality, and the classic form of military necessity.

In later years, Zvi Zamir argued the operation did reflect legitimate self-defense principles, suggesting there was intelligence information indicative of future involvement in terrorist attacks by those killed.[22] This argument was met with widespread skepticism. Herein lies one of the important questions: if there is no indication that those "targeted" are engaged in planning future acts of terrorism, whether imminent or otherwise, does this clear violation of international law make the operation ineffective? That is a significant question when we query whether effectiveness must be predicated on legality. Does effectiveness countenance illegality?

While Israeli operatives killed most—not all—of the Black September opera-tives, it did not end Palestinian terrorism. However, from the available information, that was not the goal. If it were, it would be unrealistic, reflecting a profound lack of understanding of terrorism and terrorists. On the working assumption that ter-rorism cannot be defeated (since there can be no waiving a white flag of surrender and a signing ceremony on the *USS Missouri*), the question rather is whether opera-tional counterterrorism serves as a deterrent—immediate or long-term in nature—and preventive in the face of an imminent threat.

Those questions must be added to the important recognition that Israel killed Black September members on European soil, thereby powerfully demonstrating the long arm but also violating sovereignty, a principal tenet of international law.

There is no indication Israel provided advance information, discreetly or oth-erwise, to the European countries where the attacks occurred. There is, however, no documentation of complaints filed, privately or publicly.[23] This would tend to suggest European leaders turned a blind eye, in the spirit of "understanding." This is one of the few Israeli counterterrorism operations that occurred outside the West Bank or the Gaza Strip. From the perspective of deterrence—an important measuring tool when assessing effectiveness—the "reach" is an important com-ponent, sending a clear message to terrorist organizations. However, that in and of itself does not define long-term effectiveness, regardless of impressive operational capability.

Military Necessity

The traditionally understood definition of *military necessity* focuses on whether the use of force is justified in responding to or protecting from an attack. The second part of the definition requires determining whether the use of force—if it meets the test of military necessity—is proportional to the threat and used discriminately. This analysis is intended to articulate the limits of power while recognizing that self-defense is legitimate. In examining "Operation Wrath of God," we must ask whether Prime Minister Meir's decision to kill the Black September operatives met the test of military necessity, informed by proportionality and discrimination, under the umbrella of legitimate self-defense.

Our analysis is predicated on the assumption that the intelligence information was accurate in identifying the targets deemed responsible for the attack. That does not necessarily mean the decision met the test. Rather, a narrow interpretation of military necessity focuses our analysis on the following points of consideration:

1 Did the individuals pose a continued threat to Israeli national security?
2 Did the attack, regardless of its impact or consequences, endanger Israeli na-tional security?
3 Would a failure to respond signal "weakness," thereby encouraging other terror-ist organizations?
4 Would the Operation serve as a deterrent to other terrorist organizations?
5 Would the Operation allay public morale? Is that a relevant consideration?

6 Would the Operation establish proof of the "long arm" of Israel?

As noted above, the suggestion that the targets were planning additional attacks has been largely discounted meaning our analysis of military necessity must focus on broader, more strategic considerations rather than specific individuals. From the available information, Prime Minister Meir's motivation was seemingly twofold: *revenge*, which is illegal under international law, and *deterrence* (lawful) with the former being the dominant motive.[24] Focusing on a combination of revenge and deterrence, the motivation was not to punish as the term is traditionally understood— vindication of rights and justice—but rather a combination of revenge (tactical) and deterrence (strategic).

The checklist above suggests a rationale-based approach in assessing whether a state action meets the military necessity test. While the revenge may be understandable and deterrence is an accepted argument, these do not justify the decision under military necessity. Absent reliable indication of planned future events, the Operation does not pass the military necessity test, notwithstanding that some weight can be ascribed to the deterrence rationale.

The Public

By all accounts, the Israeli public was horrified by the Munich attack, both because of the murder of 11 innocent athletes and by the location and historical context of the attack. That is very clear from reading accounts of public reaction and watching gatherings, such as funerals. The expression, "the event is seared into the collective memory" is not an exaggeration. More than that, the decision by the International Olympic Committee (IOC) to continue with the Games after a 24-hour hiatus reinforced for many Israelis the notion of being abandoned by the world, yet again. When IOC Chairman Avery Brundage, in a rambling speech announced: "The Games will go on," the Israeli public was confronted with the stark reality of being alone.

In that sense, Prime Minister Meir's decision can be understood as a galvanizing moment, reflecting an "enough is enough" attitude. We, Israelis, will be aggressive and deadly—when necessary—wherever the target may be. If this meant violating European sovereignty, so be it. The circumstances dictated such an undertaking. The combination of location and the act of Palestinian terrorism can be understood as justifying the operation.

For a public reeling from the sheer audacity and success of the attack, the decision was appropriate, justified, and much needed. While that is understandable, that is not sufficient to pass either the morality or military necessity tests proposed previously. Public sentiment notwithstanding, Prime Minister Meir's decision must, accordingly, be categorized as both immoral and illegal, regardless of the justification and necessity from her perspective. While public opinion and perception are a reality for politicians, they cannot serve as the basis for decisions of this nature. There is a direct conflict between the politician's sense of public will and the national leader's obligation to uphold the law, whether domestic or international.

Opening the door, as we shall see below, to policies reflecting a bend to public clamoring rather than adherence to the law is a path best not traveled, temptation notwithstanding.

What We Can Learn from This

The above discussion highlights the tension between legal and moral considerations. The "need" to act in response to a particular act of terrorism, in addition to the loss of innocent life, is compounded by powerful symbolism and consequences. Even under strained and challenging conditions, the nation-state must be beholden to the rule of law.

As we learned in subsequent years, an overreaction to terrorism—examples are rife in the aftermath of 9/11—is, perhaps effective in the short term but counter-productive in the long term. This is the difference between a tactical and a strategic approach to counterterrorism; the former is understandable in the moment but not effective (or smart) in the long run. The considerations are impacted when questions of morality and military necessity are introduced into the equation.

The world of 1972 is very different from that of today. In the years following the Munich Olympics, a wide array of terrorist attacks occurred domestically and internationally, with some "cross-over" such as the Boston Marathon.[25] The imposition on the nation-state to respond in accordance with the rule of law—whether domestic or international—is widely understood to be a requirement, albeit with flexibility. The tension is a reality confronted by decision-makers either confronted with intelligence information indicating a future terrorist attack or facing banner headlines screaming death counts of innocent civilians killed in a terrorist bombing. In both scenarios, decision-makers must not only make difficult operational choices but must choose whether to give weight—and if so, how much—to considerations extending beyond the "go-no go."

A particularly effective means to examine this paradigm is the torture regime authorized and implemented by the Bush Administration in the aftermath of 9/11.[26] In the aftermath of the attack, as a woefully unprepared Bush Administration began responding to the attack, President Bush signed a memo authorizing the implementation of a torture-based interrogation regime. The policy was predicated on a legal memo—the so-called Bybee Memo—drafted by John Yoo in the Office of the General Counsel.[27]

The memo is a case of legal advice and decision-making run amok in the aftermath of a terrorist attack for which there is no excuse or defense. By the three-legged analysis to which "Operation Wrath of God" is examined—effectiveness, morality, and military necessity—the Bybee Memo is a low point in American legal-political history. It is a permanent stain that time does not—must not—wash away. The Office of the General Counsel prepared a memo that violated both international and domestic law as the United States was a signatory to the Convention Against Torture.[28] Bybee and Yoo's actions led a U.S. president to break U.S. and international law. Prime Minister Meir, conversely, did not violate Israeli law, but her decision violated international legal principles of military necessity and self-defense.

In both cases, the decisions—torture-based interrogation regime and killing of enemy operatives—are understandable from a political and domestic-morale perspective. Bush and Meir were responding forcefully to traumatic events that rocked their respective nations. According to their respective perspectives they took action in the hope of restoring public confidence. Demonstrating that the strong arm of the nation-state is engaged and forceful is understandable politically.

However, while political considerations are an important aspect of operational counterterrorism decision-making, questions of morality and the obligation of acting in accordance with international law principles are no less important. Arguably, they are more important. Political considerations are relevant in the moment, whereas morality and military necessity must be understood as having long-term, strategic impact and consequences. In that sense, both Meir and Bush fell into a similar trap—Meir by all accounts bears sole responsibility whereas Bush acted in accordance with legal advice that was blatantly illegal. Their decisions reflected acquiescence to the immediate, rather than the application of more complicated and more nuanced, normative behavior reflecting respect for the rule of law.

While acting in response to the "moment" is compelling, perhaps making for good politics, it cannot be the light by which decision-makers act. In this spirit, the reaction to Munich and 9/11 (through the lens of interrogation) reflects a failure to adhere to operational decision-making guided by respect for the rule of law and obligations of morality. Short-term operational success must be held up to the harsh glare of strategic thinking predicated on a model extending beyond the immediate. In that sense, Prime Minister Meir's decision based on what is "good in the moment" is not necessarily right in the long run, much less legal. The combination of right (morality) and legal (military necessity) outweigh considerations of effectiveness, whether short- or long-term. Otherwise, state power will not reflect principles and values in accordance with international and domestic law. The consequences for such an approach—actually a failing—far outweigh transient success, no matter how compelling and operationally impressive.

For that reason, "Operation Wrath of God," when held up to the bright lights of morality and military necessity, falls short, notwithstanding the understandable reason for which it was undertaken. The depth of pain responded to must not be used as the barometer by which the operation, or others in its wake, are measured.

Notes

1 Professor of Law, SJ Quinney College of Law, University of Utah; with thanks to Cait McKee (SJQ JD expected 2024) and Austin Sork (SJQ JD expected 2024) for much appreciated editorial assistance.
2 Yair Galily, "'We Can Only Trust Ourselves': 'Operation Wrath of God' in Perspective," *Israel Affairs* 28, no. 4 (June 15, 2022): 589–596.
3 Oroub Al Abed, *FMO Research Guide: Palestinian Refugees in Jordan* (Refugee Studies Centre, 2004).
4 https://www.ushmm.org/collections/bibliography/the-eichmann-trial
5 Ibid.
6 Ibid.
7 Galily, 'We Can Only Trust Ourselves.'

8 Alexander B. Callahan, *Countering Terrorism: The Israeli Response to the 1972 Munich Olympic Massacre and the Development of Independent Covert Action Teams*, PhD diss., Marine Corps Command and Staff College, 1995.
9 Ibid., at 8.
10 Ibid.
11 Galily, 'We Can Only Trust Ourselves.'
12 Ibid.
13 Ibid.
14 David Carlton, "Rewarding the Palestinians," in *The West's Road to 9/11: Resisting, Appeasing and Encouraging Terrorism since 1970* (London: Palgrave MacMillan, 2005), 47–63. https://doi.org/10.1057/9780230508767.
15 Galily, 'We Can Only Trust Ourselves.'
16 Ibid.
17 Ibid.
18 A. Silke and A. Filippidou, "What Drives Terrorist Innovation? Lessons from Black September and Munich 1972," *Security Journal* 33 (2020): 210–227.
19 Michael Rubner, "Massacre in Munich: The Manhunt for the Killers behind the 1972 Olympics Massacre/One Day in September: The Full Story of the 1972 Munich Olympics Massacre and the Israeli Revenge "Operation Wrath of God" Striking Back: The 1972 Munich Olympics Massacre and Israel's Deadly Response/Vengeance: The True Story of an Israeli Counter-Terrorist Team," *Middle East Policy* 13, no. 2 (2006): 176.
20 Joan Fitzpatrick, "Speaking Law to Power: The War against Terrorism and Human Rights," *European Journal of International Law* 14, no. 2 (2003): 241–264.
21 Kim Cragin and Scott Gerwehr, *Dissuading Terror: Strategic Influence and the Struggle against Terrorism* (Santa Monica. CA: Rand Corporation, 2005).
22 Galily, 'We Can Only Trust Ourselves.'
23 Ibid.
24 Callahan, *Countering Terrorism.*
25 Silke, "What Drives Terrorist Innovation? Lessons from Black September and Munich 1972."
26 Kathleen Clark, "Ethical Issues Raised by the OLC Torture Memorandum," *Journal of National Security Law & Policy* 1 (2005): 455.
27 Ibid.
28 Ibid.

11 Military Necessity in the Gray Zone

Joshua Hastey

Introduction

In 2014, hundreds then thousands of armed men in military uniforms with no national insignia appeared in the Ukrainian region of Crimea. Moving in organized, coordinated maneuvers, these soldiers were widely believed to be operating either at the behest of or as agents of the Russian government, but the confusion and hesitation created by the ambiguous origin of these forces delayed Ukrainian and international responses enough to allow the "green men" to establish a *fait accompli* before any meaningful resistance could be mounted. In 2016, hackers attacked the Central Bank of Bangladesh, targeting over $1 Billion and successfully transferring over $100 million. Though many now believe the heist was carried out by the North Korean government, attribution was difficult and enforcing any repercussions remains challenging. In 2020, Indian border patrol units in the Himalayas discovered Chinese army units patrolling and encamped within Indian claimed land, well beyond what had previously been the effective line of control dividing the two states' disputed border region. For the two years that followed, Indian forces scrambled to halt creeping, progressive encroachments along the line of actual control, fearing a series of "salami-slicing" *faits accomplis*. Each of these represents a form of conflict pervasive in international politics, both today and throughout history, which scholars have come to call gray zone operations. Though more common, this form of conflict has received less attention than conventional war from scholars of international relations and the just war tradition. However, as I argue below, this volume's renewed attention to the military necessity standard highlights two complementary framings of military necessity—restriction and stewardship—that help provide moral clarity in the strategic ambiguity of the gray zone.

Because the just war tradition has dedicated relatively little attention to the gray zone between conflict and cooperation, and most of that attention has sought to branch off a distinct paradigm of *jus ad vim*, this chapter needs to address two preliminaries at the outset: (1) "Why use a realm of interstate conflict that falls outside of conventional warfare to expound on a standard in a tradition that historically addresses conventional warfare?"; and (2) How do we conceptualize gray zone operations in a way that reaches beyond conventional warfare but does not include

DOI: 10.4324/9781003390398-11

the whole realm of statecraft. The remainder of this chapter proceeds in four parts. The first two sections provide foundational responses to the two questions above, beginning with a justification for exploring military necessity from the perspective of gray zone operations and proceeding to offer a working conceptualization of gray zone activity. The third section builds on insights into the military necessity doctrine from O'Brien and Patterson to suggest a framework for understanding military necessity as vital to extending the moral clarity provided by the just war tradition to the shadows of the gray zone.

Why Apply Military Necessity to Gray Zone Operations?

If conflict in the gray zone explicitly refers to something other than conventional warfare and the just war tradition seeks to provide moral clarity in warfare, why would we use gray zone operations as an avenue to explore the military necessity doctrine? In short, the answer is twofold. First, conflict in the gray zone is the most common operational vector of interstate competition. Second, the military necessity doctrine is the linchpin holding together the *ad bellum* standards of right intent and last resort with *in bello* standards of distinction and proportionality. The troublesome, conceptually ambiguous realm of gray zone operations counterintuitively provides an excellent context to illustrate this important role of the military necessity doctrine.

Conventional thinking about war and peace do not capture the full breadth of armed and coercive conflict among states. In fact, as research by a growing cadre of political scientists has shown, the bulk of the coercive use of force— from unilateral military revisions to territorial status quos to operations in the space, cyber, and cognitive domains—takes place outside the conventional definition of war as an armed conflict resulting in 1,000 or more battle deaths per year.[1]

For scholars in the just war tradition, the gray zone demands our attention and challenges our analytical paradigm of war. Operations in the gray zone demand our attention because they are replete throughout the international system and because their use or misuse has important effects on the establishment and maintenance of an Augustinian justly ordered peace. At the same time, gray zone operations challenge our paradigm of war by belying the continuous, not dichotomous, nature of war and peace. In international politics, conflict is not a binary toggling between war and peace but a continuum, and that continuum is primarily characterized by probing, signaling, and skirmishing that is certainly conflictual but does not escalate to what we commonly call war. Still, these behaviors are not novel, however sophisticated the tools they employ. To paraphrase LiVecche's logic, attacks in the gray zone are attacks nonetheless, and thus the wisdom of the just war tradition is equipped to address them.[2]

The realization that conflict between war and peace is in need of a framework to provide moral clarity is not new. As early as 2006, Walzer argued that we "urgently need a theory of just and unjust uses of force."[3] Recognizing this, several scholars have suggested a novel paradigm, *jus ad vim*, to address force short of war.[4] However, as a colleague and I have argued elsewhere:

> …the introduction of *jus ad vim* as an alternate framework to *jus ad bellum* and *jus in bello* misunderstands the continuity of the spectrum of competition as an unbroken continuum of conflict. Second, a collateral casualty of the introduction of *jus ad vim* as an alternate framework to *jus ad bellum* has been a neglect of the extension of *jus in bello* considerations to gray zone conflict. Returning focus to the holistic *bellum justum* framework returns the integration of *ad bellum* and *in bello*, allowing both sets of principles to help clarify the moral ambiguities of the gray zone.[5]

Indeed, this volume's reintroduction of military necessity only furthers the insights the just war tradition can provide to scholars and policymakers operating in the gray zone.

Conceptualizing the Gray Zone

But what is the so-called 'gray zone'? The term refers to an expansive set of foreign policy tools that seek to realize a strategic advantage over an adversary without provoking outright war. In other words, it is a range of conflictual behaviors that aim to gain the benefits of a favorable change to the status quo without the costs of fighting the defender of the status quo. Such strategies rely on ambiguity and asymmetries of interest and access to make gradual status quo revisions that defenders of the status quo are disincentivized to resist.

Among the classical war theorists, we are reminded again and again of the continuity of coercive efforts from information operations to isolated conflict and maneuver to massed conventional war. Sun Tzu writes that the "acme of skill" is to "win without fighting" in referencing the use of coercive force and diplomatic maneuvering to induce surrender.[6] Clausewitz notes that war is merely the "continuation of political intercourse by other means" in noting the nature of war as an ongoing bargaining exercise.[7] Mao and Galula repeatedly underscore the reciprocal nature of information operations and material advantage in fighting indigenous and proxy insurgencies, while Corbett and Mahan develop theories of naval coercion short of war and guidance for identifying the time to escalate and pursue force-on-force engagement.[8] These classical theorists are met by current military and strategic doctrine. To note only one example of many, American joint doctrine explicitly identifies force short of war as but one region on a continuous spectrum, across which policy can shift back and forth.[9] The fundamental logic of gray zone operations is straightforward: use the range of a state's elements of national power—diplomatic, informational, military, and economic tools—to coercive effect against adversaries. Importantly, the key feature that distinguishes this sort of coercion from conventional war and hybrid war is its intent: gray zone operations seek to coerce *without* escalating or provoking an adversary to escalate to conventional war.

If that is the logic, how are we to recognize gray zone operations when we see them? Luckily, the security studies literature has built an initial foundation. Javier Jordán articulates four distinguishing marks of gray zone strategies: ambiguity, multidimensionality, asymmetry of interests, and gradualism, which I will briefly summarize in turn.[10] Ambiguity refers to two related characteristics of gray zone operations. First it refers to the challenge of distinguishing conflictual gray zone behavior from peaceful competition on the one hand and low-level armed conflict on the other. Second, it refers to the frequent difficulty in attributing gray zone aggression to specific actors. Though sometimes attribution is clear, one common tactic in gray zone operations is to obfuscate aggressors' identity, even if only temporarily, to delay or deter effective responses. We see this most commonly manifested in cyber-attacks where attackers deliberately mask their identity in an effort to make it more difficult or impossible for their targets to respond effectively. Multidimensionality refers to the incorporation of operations and tactics across a range of contexts from social and political influence to information and disinformation campaigns to the coercive use of military and economic instruments of national power. Asymmetry of interests highlights a state's ability to leverage relative differences in interest in a given region or issue area between attacker and target. That is to say, when a contentious issue is of more value to the attacker than the defender the defender is less likely to commit great resources and attention to the issue than the attacker is. Finally, gradualism refers to instigators ability to calibrate the level of pressure they bring to bear such that they make progress in increments just low enough to avoid triggering strong responses from their targets.

Military Necessity: Stewarding Capabilities in the Uncertainty of the Gray Zone

In his canonical treatise *On War*, Clausewitz characterizes war as, at its best, full of "fog and friction," making even the best laid plans subject to uncertainty and chance.[11] As we have seen, operations in the gray zone are designed to magnify the fog, friction, and uncertainties associated with conventional conflict in order to cut off avenues of response to current and would-be adversaries. The same characteristics that make gray zone operations challenging to respond to at the strategic and operational levels also complicate attempts to bring moral clarity to bear. The ambiguity of actions in the gray zone, the incremental nature of gradual revisions and the studied avoidance of giving a target clear *causus belli* present a real challenge to strategists and ethicists alike.

Helpfully, previous scholars in the just war tradition have already raised two key insights into the military necessity doctrine that can help us navigate the complexity of the gray zone. The first, noted by William O'Brien, is that military necessity helps focus and restrain the use of force within conflicts by serving as a lens to evaluate proportionality.[12] This framework limits the use of force to operations which are likely to bring the state closer to achieving its right intent, and interprets the proportionality of potential harms inflicted by the operation in light of the furtherance of the stated objective. Thus, in the restrictive sense, military necessity is rightly understood as a proscriptive doctrine: "Action *A* under consideration may not be undertaken unless it can be reasonably expected to raise the chance of victory or to minimize risk of harm to noncombatants."[13] In this framing, we see reflected attention to two things: (1) Achieving the just aims of the war through victory; and (2) Minimizing harm to noncombatants.

The second framing maintains attention to drawing the *ad bellum* right intent into *in bello* considerations, but it differs in its second consideration. Patterson highlights the role of military necessity in challenging leaders to steward the lives of their soldiers and their military capabilities.[14] This is important for at least two reasons. First, it recognizes that soldiers do not entirely abdicate their rights to protection. As a category, soldiers do accept greater risk so that noncombatants can be spared but neighbor love still compels soldiers to look out for their brothers in arms and commanders to care for the lives entrusted to them. Second, the doctrine of military necessity puts into practice what Patterson has referred to as the morality of victory.[15] This follows from the *ad bellum* consideration of right intent. A legitimate authority wielding the sword with just cause and right intent has a moral duty to achieve the right intent aimed for at the outset of the war. Viewed from this perspective, the military necessity doctrine is a prudential extension of both deontological duties: love neighbor and preserve a just order. Thus, military necessity acts both as a restraint against wonton destruction *and* a practical means of focusing the use of force to minimize risk to the overall mission and maximize the chances of succeeding in attaining the *ad bellum* intent.

Illustrating the Reasoning: Military Necessity in Maritime Gray Zone Operations

In 2009, as part of an escalating set of provocations aimed at establishing increased de facto control in the South China Sea, armed Chinese vessels and aircraft harassed and used water canons to chase off two unarmed U.S. Navy survey ships—the USNS Impeccable and the USNS Victorious—conducting routine survey missions well outside the 12- and 24-nautical-mile boundaries that mark Territorial Seas and Contiguous Zones.[16] In 2011, Chinese paramilitary and China Marine Surveillance (CMS) forces clashed at least ten times with Vietnamese and Philippine civilian vessels operating in or near contested South China Sea (SCS) waters.[17] A review of conflicts between states in the region over the past two decades indicates that the end of the 2000s reveals sharp rise in the number of incidents between China and other SCS claimant states, culminating in the seizure of a series of maritime features in the Spratly Islands and development of these features into fortified concrete garrisons scattered well into what is internationally recognized as Filipino and Vietnamese waters. These events include assertive actions taken by the People's Liberation Army Navy (PLAN) of China, the CMS, and Chinese fishing vessels operating independently but with the blessing of PLAN security forces.[18] This pattern of increasing confrontations, strengthening maritime forces, and fortifying its held features in the region in order to better defend its claims typifies the key elements of aggressive operations in the gray zone: they leverage ambiguity in the use of dispersed and dissociated forces to achieve favorable revisions, concentrated interests in the face of distracted or divided opponents to achieve interest asymmetry, and move gradually across multiple dimensions to diffuse possible responses to aggressive operations.

However, in the East China Sea (ECS), we observe a different pattern. Even as China's forces were beginning its revisionist behavior in the SCS, and maritime self-defense force was preparing to deter similar behavior in the ECS through a series of gray zone and diplomatic operations of its own. A brief survey of these operations shows a focus on building resilience to PRC encroachment and providing credible deterrence against efforts to seize control of maritime territory in the ECS.

The first prong of this effort has been Japan's commitment to hardening its defenses of disputed islands and waters through force, establishing a regular presence, and building redundant platforms to be able to respond to incursions effectively. To highlight just a few instances of this hardening, over the past decade, Japan has committed tremendous additional resources to its Coast Guard, including the creation in 2016 of a 12-ship unit dedicated specifically to patrolling the waters around Senkaku (a key disputed island in the ECS), a budget increase to $1.87 billion in 2017, and the deployment of persistent intelligence, surveillance, and reconnaissance assets with real-time video feeds of disputed waters in 2018 and 2019.[19] Japan's regular patrols, combined with its physical presence on the disputed islands, means that a successful Chinese *fait accompli* in the region would have to expel existing military and paramilitary forces at a minimum, and would risk escalation to an outright war at a maximum.[20]

As part of its diplomatic efforts to deter continued PRC incursions into the ECS, Japanese diplomats elicited a series of statements of support from high level American officials, starting with statement from the US State Department affirming that the Senkaku islands are subject to the US-Japan alliance treaty.[21] This statement was later reaffirmed in a statement by President Barack Obama in 2014, in which he stated,

> The policy of the United States is clear—the Senkaku Islands are administered by Japan and therefore fall within the scope of Article 5 of the U.S.-Japan Treaty of Mutual Cooperation and Security. And we oppose any unilateral attempts to undermine Japan's administration of these islands.[22]

These statements, while in keeping with longstanding American policy toward Japan, were significant in that they were public commitments to Japan, and signals to China, that the United States was committed to preserving the status quo in the ECS. This public commitment served not only to reassure Japan of US commitments, but to raise the reputational costs to China of attempting to revise the status quo in the ECS.

In Japan's responses to China's aggressive gray zone operations in the ECS, we have seen that Japan has been able to raise China's costs of imposing *faits accomplis* materially, through its regular patrols through disputed waters and garrisoning of Senkaku, and nonmaterially, through diplomatic maneuvering to extract public support of its position in the dispute from the United States.

Here we see both the strategic logic of opposing gray zone operations through hardening targets and presenting credible deterrence, but we also see the prudential and moral logic of the military necessity standard at work. By outmaneuvering China in repeated crises in the ECS, Japan succeeded in preventing Beijing from imposing major revisions to the maritime status quo in the region, but it also did so while preventing escalation and the potentially catastrophic effects they could have on servicemen and noncombatants alike. It's admittedly difficult to tease out the distinction between actions with both strategic and moral benefits but it is worth noting that, at least in this case, there is a high degree of congruence between successful opposition of postal actions in the gray zone and careful attention to the military necessity standard.

Conclusion and Need for Further Work

In this chapter, I have argued that the military necessity doctrine points the just warrior toward the two complementary principles of restraint and stewardship. Military necessity joins proportionality and distinction in restraining the tendencies of war, focusing the use of force on operations that are most likely to further the right intent found in the *ad bellum* considerations. The stewardship principle calls military leaders both to neighbor love for the soldiers under their command and to the calculated focus of the military resources at their disposal at the key decisive points of battle. Unlike the restraint principle, the stewardship principle is primarily directed at the judicious conservation of capabilities in order to be able to apply maximum pressure at the moment it will be most effective. Importantly this may not always mean pitched battle. In fact, as the cases examined in this chapter illustrate, the conflict and application of force may be quite low intensity. But the stakes may nonetheless be high, and the pursuit of a more just and more ordered peace may be merited.

I hope this brief exploration serves to start a broader conversation on the moral clarity that the historical *jus ad bellum* and *jus in bello* standards can offer to strategists and warriors operating in the gray zone. There are hard historical cases—such as the bombing of Hiroshima analyzed elsewhere in this volume—to which the military necessity standard can provide important clarity. There are fascinating and badly needed explorations of how the just war tradition can illumine problems of deterrence, diplomatic signaling, operations in the cognitive domain, and economic coercion waiting to be written, each of which would be important in its own right and would contribute to a body of work providing moral guidance in the most common domain of international conflict.

Notes

1 Michael J. Mazarr, "Mastering the Gray Zone: Understanding a Changing Era of Conflict," (US Army War College Strategic Studies Institute Carlisle, 2015); Dan Altman, "By Fait Accompli, Not Coercion: How States Wrest Territory from Their Adversaries," *International Studies Quarterly* 61, no. 4 (2017): 881–891, https://doi.org/10.1093/isq/sqx049; Joshua Hastey and Adam Knight, "New Under the Sun? Reframing the Gray

Zone in International Security," *Journal of Strategic Security* 14, no. 4 (2021): 21–36; Joshua Hastey, *China, Faits Accomplis and the Contest for East Asia: The Shadow of Shifting Power* (Abingdon: Routledge, 2023).

2 Marc Livecche, "Just War & Cybersecurity: The Old World Can Help Us with the New," *Providence* June 24, 2021.

3 Michael Walzer, *Just and Unjust Wars: A Moral Argument with Historical Illustrations*, 4th ed. (New York: Basic Books, 2006), 106.

4 Daniel Brunstetter and Megan Braun, "From jus ad bellum to jus ad vim: Recalibrating Our Understanding of the Moral Use of Force," *Ethics & International Affairs* 27, no. 1 (2013): 87–106; Jai Galliott, *Force Short of War in Modern Conflict: Jus ad Vim* (Edinburgh: Edinburgh University Press, 2019).

5 Joshua Hastey and Adam Knight, "Just War Standards in Shades of Gray," IS-ISSS Conference Presentation, Gainesville, FL 2022.

6 Sun Tzu, *The Art of War*. Translated by Samuel Griffith (London: Oxford University Press, 1971).

7 Carl von Clausewitz, *On War*. Edited and Translated by Michael Howard and Peter Paret (Princeton, NJ: Princeton University Press, 1989).

8 Mao Tse-tung, "On Protracted War," May-June 1938; M. I. Handel, *Masters of War: Classical Strategic Thought* (Mahan, AT: Routledge, 2009). *On the Influence of Sea Power Upon History: 1660–1783* Sagamore Press (1957); Sir Julian Corbett, *Some Principles of Maritime Strategy* (Annapolis, MD: Naval Institute Press, 1988).

9 Joint Publication 3-0: Joint Operations (2017); Joint Doctrine Note 1-19: Competition Continuum (2019).

10 Javier Jordán, "International Competition Below the Threshold of War: Toward a Theory of Grey Zone Conflict," *Journal of Strategic Security* 14, no. 1 (2020): 2–4, https://doi.org/10.5038/1944-0472.14.1.1836.

11 von Clausewitz, *On War*.

12 William O'Brien, *The Conduct of a Just and Limited War* (New York: Praeger Publishers, 1981), 85.

13 Though the reasoning varies, this is consistent with the general framing of military necessity in the LOAC literature. See Y. Dinstein, *The Conduct of Hostilities under the Law of International Armed Conflict* (Cambridge: Cambridge University Press, 2004); Y. Beer, *Military Professionalism and Humanitarian Law: The Struggle to Reduce the Hazards of War* (Oxford: Oxford University Press, 2018); J. D. Ohlin and L. May, *Necessity in International Law* (New York, NY: Oxford University Press, 2016).

14 Eric Patterson, "Introduction: Returning Military Necessity to the Jus in Bello," in Eric Patterson and Marc Livecche, eds., *Returning Military Necessity to Just War Statecraft* (London and New York: Routledge, 2023).

15 Eric Patterson, *Just American Wars: Ethical Dilemmas in US Military History* (London: Routledge, 2018).

16 Ronald O'Rourke, *Maritime Territorial and Exclusive Economic Zone (EEZ) Disputes Involving China: Issues for Congress* (Washington D.C.: Congressional Research Service, 2012).

17 O'Rourke, *Maritime Territorial and Exclusive Economic Zone (EEZ) Disputes Involving China*, 13–17.

18 For example, see: Aileen Baviera, *Deference/Defiance: How Does the Philippines Cope with China? The Domestic Politics Dimension* (Manila, 2012); "Stirring up the South China Sea: Regional Responses," (International Crisis Group, 2012); Carlyle Thayer, "South China Sea Disputes: Asean and China" (2011); "Japan Protests China's Gas Development in East China Sea," *The Asahi Shimbun*, September 15, 2015; Jane Perlez, "China Said To Turn Reef into Airstrip in Disputed Water," *New York Times*, November 23, 2014; David Sanger and Rick Gladstone, "Piling Sand in a Disputed Sea, China Literally Gains Ground," *New York Times*, April 8, 2015.

19 Japan's Ministry of Defense. "Defense of Japan 2019," 220-222, https://www.mod. go.jp/en/publ/w_paper/wp2019/pdf/index.html; Edmund J. Burke, et al., *China's Military Activities in the East China Sea: Implications for Japan's Air Self-Defense Force*, 10–11.
20 Sung K. Hyong and David A. Anderson, "Japan's Security Strategy and Its Impact on U.S. National Security Interests," *Inter Agency Journal* 10, no. 1 (2019): 3–33; Koichi Sato, "The Senkaku Islands Dispute: Four Reasons of the Chinese Offensive—A Japanese View," *Journal of Contemporary East Asia Studies* 8, no. 1 (2019): 50–82; Yun Yu and Ji Young Kim, "The Stability of Proximity: The Resilience of Sino-Japanese Relations over the Senkaku/Diaoyu Dispute," *International Relations of the Asia-Pacific* 19, no. 2 (2019): 327–355.
21 Glenn D. Hook, Julie Gilson, Christopher W. Hughes, Hugo Dobson, *Japan's International Relations: Politics, Economics and Security*, 3rd ed. (Oxon and New York: Routledge, 2012).
22 Ankit Panda, "Obama: Senkakus Covered under US-Japan Security Treaty," *The Diplomat*, April 24, 2014.

12 Military Necessity as Moral Imperative

Just War and Hiroshima

Marc LiVecche

Moral Horror

On August 6th, 1945, a specially modified U.S. Army Air Force B-29 "Superfortress" heavy bomber, the *Enola Gay*, cruised some 31,000 feet above Honshu, the largest of the Empire of Japan's four main home islands. Conditions were good: mostly clear skies and limited cloud cover. The *Enola Gay* had no trouble locating her target. At 08:15 AM, her bombardier released a 9,700-pound atomic bomb, christened "Little Boy," over the port city of Hiroshima. Decorated with crude messages for Emperor Hirohito, the device fell for 43 seconds. A crosswind pushed at it, causing it to deviate slightly, just off bullseye. At 1,900 feet above Hiroshima, several thousand pounds of conventional explosives ignited inside Little Boy discharging a canon-like mechanism that fired a plug of one subcritical mass of uranium along a barrel and into a second—hollowed-out—subcritical mass of uranium. The resulting compression and increased mass forced the two pieces of uranium to become a single, supercritical mass, setting of a fission chain reaction that continued until the energy released became so great that the bomb simply blew itself apart. Little Boy detonated with a force greater than 12,000–15,000 tons of TNT.

The temperature at the blast's burst point eclipsed a million degrees Celsius and ignited the air surrounding it, resulting in a fireball some 840 feet in diameter with an apparent brilliance ten times the brightness of the sun and a temperature hotter than the sun's own surface. The blast wave shattered windows over a distance of 10 miles and was felt at a distance of over 37 miles. With a destruction radius of 1 mile, the thermal pulse sent fires raging over 4.5 miles. People on the ground reported a brilliant flash, a strange smell, and a booming noise. The city toppled, buildings were ripped from their foundations, bridges twisted, some 70 percent of the city's structures were shoved to pieces.[1] Radiant heat traveling at the speed of light caused flash burns, charring skin to charcoal. Probably somewhere between 70,000 and 80,000 souls were consumed instantly. Tens of thousands more would die more slowly, succumbing to injuries or radiation sickness, in varying degrees of agony, days or weeks, or even years after. About one-half-hour later after the initial blast, a black rain fell from the darkened heavens. A stew of dirt, ash, and radioactive particles that were sucked up into the air with the mushroom cloud, the rain

DOI: 10.4324/9781003390398-12

poisoned areas initially uncontaminated by the detonation. Hiroshima had become "a place of desolation."[2]

Three days later, the Empire of Japan accepted Allied demands for unconditional surrender.[3]

The bombing of Hiroshima was, as the section header proclaims, a moral horror. But to avoid misunderstanding, it's important to note that the stress is meant to be placed on both terms. For certain, it was a moral *horror.* It rightly shocks the conscience. All things being equal, we ought to prefer that it was never done. However, all things in the waning summer of 1945 were not equal and, given the manner in which they were not, while the bombing was morally shocking, it was also morally right—it was a *moral* horror. To sharpen the point—and the discomfort—further: to not have dropped that bomb would have been morally wrong.

How can this be?

There is a tension in war that is sometimes framed as the dialectic between military objectives and humanitarian or moral principles. While it's obviously true—even a *truism*—that an *illegitimate* military objective is rightly in tension with moral concerns, the tension isn't resolved simply by pursuing an objective that *is* legitimate. For it may be, as some insist regarding the link between the Hiroshima bombing and Japanese surrender, that a proper objective can be pursued through unjustified means. And yet, the tension is still not necessarily resolved even if we have both a legitimate military objective and a justified means of pursuing it. One reason for this is because there is a lot riding on what the term "justified" means. Some argue that simply because context and limited options might mean particular kinds of horrible actions are temporarily justified, this still doesn't make them morally right. One must simply live with the moral tension of overruling moral rules if the ends are critical enough.

In ways similar to J. Daryl Charles' assertion in his early chapter in this volume that humanitarian and military necessity are not polar opposites, this paper contends that military necessity and moral concern can live side-by-side without contradiction. It goes further to suggest that military necessity can, in particular circumstances, help shape the moral analysis of a conflict situation. What it doesn't endorse is the proposal that military necessity can simply abrogate moral norms. Given this volume's proposition, therefore, if the military necessity doctrine is promoted from a tacit to an articulated element of just war statecraft, the doctrine of necessity must distance itself from the logic of supreme emergency. Military necessity must serve, not sideline, the just war framework in its function to help direct the moral use of proportionate and discriminate force, including in hard cases, to appropriately intended aims.

One benefit of this is that under this scheme military necessity will prove itself to be both a restraint as well as a spur to action. This is a remedial boon. Not only does it help rectify the tendency of too many thinkers who view the just war framework as simply a leash on force, but it realigns the *jus in bello* with the *jus ad bellum* pattern, which has always had both. Proper authority, just cause, and right intent, viewed as the deontological elements of the framework, alert the moral analyst that force is permitted or perhaps even required, while the prudential elements—last resort, proportionality, and probability of success caution restraint. Just so, necessity serves as a catalyst while proportionality and discrimination serve as cautions.

To offer a preliminary sketch of what this might look like, this chapter offers a partial just war defense of the bombing of Hiroshima on the grounds of necessity. To narrow the scope, it does so by defending two key assertions: First, the Allied demand for unconditional surrender by the Japanese was a morally appropriate—I will assert *required*—war aim. Second, the atomic attack offered the best means to secure this unconditional surrender. For reasons that will be clear downstream, I will add qualifiers to each of these assertions. In aggregate: the atomic bombing of Hiroshima offered the most *moral*—that's to say, *the most proportionate and discriminate*—means possible to obtain the unconditional surrender of the Japanese *in the quickest possible time*. In the summer of 1945, these qualified assertions were not merely morally appropriate, they were morally essential. Why they were so will be touched upon, again only briefly, in conclusion.

The Just War Proposal: Necessity and Morality Together Again

As even a cursory assessment of the scholarship, including the contents of this volume, makes clear, beyond a rough agreement that particular military situations seem to require, or *necessitate*,[4] particular military responses, the notion of military necessity—and its relationship to the just war moral framework—regularly means different things to different thinkers. Consequently, there exist sometimes vastly different normative evaluations of necessity.

The American protestant theologian Reinhold Niebuhr, for example, is probably best known as the steadfast defender of democracy against the totalitarian evils of the twentieth century's fascist and communist regimes. Indeed, Niebuhr came to increased national prominence in the lead up to the Second World War by making the case for American intervention against Nazism. To help his doing so, he inaugurated a new publication, *Christianity & Crisis*, and, shortly after Japanese Zeros dropped from the December skies over Pearl Harbor, he insisted in its pages that the Christian faith offers "no easy escape from the hard and sometimes cruel choices of such a world as ours; but that it did offer resources and insights by which our decisions could be made wisely and our responsibilities borne courageously." Niebuhr suggested that it was to America's own good that the Japanese attack had finally "strengthened our reluctant will and overruled our recalcitrant will," goading America to now do what she ought already to have done. "We have been thrown into a community of common responsibility," Niebuhr suggested, "by being engulfed in a community of common sorrow."[5]

While Niebuhr might have rejoiced over the moral rousing of American power, he did not rejoice in its need to be roused. However necessary, Niebuhr cautioned that war had terrible costs, even theological ones, including a necessary renunciation, if partial, of the ethics of Christ. This worry comes from the tension—contradiction even—that Niebuhr saw between the requirements of love and those of justice. This tension results in a paradox which Niebuhr might have summarized as follows. The moral vision of the New Testament, specifically as revealed in the life of Christ, declares the Law of Love to be the normative ideal for Christian behavior. Given the conditions of history, however, this norm is impossible to follow.

Alongside the Impossible Ideal is the possibility of approximating those ideals. Given these options, in the face of sufficiently grave political evil, the Law of Love requires that we overrule love.[6] The theological costs Niebuhr ascribed to paradox can be summed in a single phrase: "It is not possible to move in history without becoming tainted with guilt."[7]

This dialectic between love and justice should remind us of the supposed dialectic between military necessity and moral principle. It parallels, too, the common view within just war thinking that military necessity is essentially the same thing as "supreme emergency." Most often, supreme emergency appears to cast necessity as a pause—or hiatus—from just war morality. As Daniel Bell describes it, this view suggests that one follows the just war criteria until the costs become too high. At that point, supreme emergency provides a kind of get-out-of-jail free card overruling the normative restrictions that overly-constrain the warfighter in a particularly high-stakes crisis.[8]

Michael Walzer's advancement of supreme emergency in *Just and Unjust Wars* and, perhaps more especially, his subsequent clarification in *Arguing About War*, ratified its acceptance among many just war scholars. Said succinctly, Walzer avers there are occasions in which it becomes right to do wrong. These occasions are limited. The laws of war must be obeyed "until the heavens fall."[9] The simple fact that a nation is about to face defeat is not sufficient.[10] Rather, a state of supreme emergency would only exist if the danger faced was both imminent and disastrous. There is something obviously different about the prospect of the United Kingdom losing its war against Nazi Germany versus losing its war over the Falklands to Argentina.

When the conditions for a supreme emergency do appear at hand, Walzer views the prospect in language strikingly similar to Niebuhr:

> There are moments in human history that are not governed by moral rules; the human world is a world of limitations, and moral limits are never suspended—the way we might, for example, suspend *habeas corpus* in a time of civil war. But there are moments when the rules can be and perhaps have to be overridden. They have to be overridden precisely because they have not been suspended. And overriding the rules leaves guilt behind, as a recognition of the enormity of what we have done and a commitment not to make our actions into an easy precedent for the future.[11]

Walzer tacitly acknowledges his proximity to Niebuhrian language when he summarizes that in cases of true supreme emergency, the given immorality is "no less immoral" but that it is "simultaneously, morally defensible."[12] This cousin-logic of paradox and supreme emergency places a heavy burden on just warfighters and those who send them into war. It demands that they willfully dirty their hands—and thereby their souls—in order to shoulder the martial duty to do what ought not to be done. As the Frenches made clear in their chapter, especially in light of our deepening understanding of moral injury as a spiritual wound coming from doing or allowing to be done something that goes against—or *appears* to go against—a deeply held moral norm, the burden placed on warfighters by such Niebuhrian moral analysis is a catastrophe.[13] And an unnecessary one.

This chapter contends that casting a military necessity doctrine in this way is a misreading of both necessity and just war morality. As for just war, the tradition was always intended to handle precisely these kinds of moral dilemmas without having to step outside itself. This is morally fundamental. In this chapter, I presuppose an understanding of the just war tradition that grounds the tradition's moral framework in neighbor love, which is *a*, if not *the* essential norm of Christian morality. I stand in good company. Before I came along, Thomas Aquinas, following Augustine's emphasis on love, placed his own discussion of just war in the *Summa Theologica* in the midst of his treatment of *caritas*, or charity. Not incidentally, Thomas might also have taken inspiration from the Apostle Paul, who, in his "Epistle to the Romans" (12:9), implores the Christian to "love without hypocrisy," and, in the immediate verses following, explains what this means. A bit later (13:10), Paul proclaims that "Love does no wrong to the neighbor." Inbetween these points, he discusses human government and affirms that God has ordained political authority to use "the sword" to curb wrongdoing. In both this Pauline and Thomistic bookending, then, deliberation on the use of force occurs in the midst of a discussion of love.

But as a norm of behavior, the imperative to *love* isn't always terribly clarifying about what to do next; especially when moral goods are in conflict and it seems impossible to meet the requirements of two or more clashing moral duties at the same time. For instance, firm in my understanding of the grounding norm of Christian ethics, I do not doubt that I am supposed to love my neighbor, whom I've just encountered on a darkened downtown street. But how precisely do I love that neighbor when I discover he is unjustly kicking apart the face of another neighbor and will not stop? I know I am supposed to love both of them. I also know it's not sufficient to say I'll love one of them—let's say the victim-neighbor—*now* and love the other one—let's call him the enemy-neighbor—later. I have to love both of them this very instant. But it is also clear that I cannot love both of them in exactly the same way in exactly the same moment.

The Frenches do important work in their chapter reflecting on the complexity of both loving one's enemy and killing them. Crucially, they note that some form of dehumanization is often necessary to allow for the requirements of combat. This is nothing new—surgeons and other medical professionals also employ distancing techniques. They are right to assert that while empathy might have to be temporarily "dampened" in combat, we ought never to allow it to be fully extinguished.

In supplying the resources needed to navigate such complexities, the just war moral framework reveals itself to be essentially casuistic. Casuistry has gotten a bad rap because, like much else, it has sometimes been misused in order to justify moral laxity and deny the rigor—or even existence—of moral rules. But this is untrue. Rather, casuistry, we shouldn't be too embarrassed to recall, is simply a case-based method of moral reasoning attending to the application, or interpretation, of moral norms in relation to particular ethical cases and the role of rules or principles as mediating agents. Moral reasoning requires moral judgment, judgments that are faithful to moral principles even if they do not straightforwardly derive from them. The casuist simply questions how to interpret and apply the rule in light of context, particularly in those scenarios in which moral goods appear to be in conflict. The casuist—analyzing the conflict situation before him, comparing it to normative paradigms, identifying relevant presumptions, assessing the details of context, and referring to reason, authority, and experience—attempts to move from the general duty to the specific manifestation of that duty in the present moment—seeking to know how the given rule regulates moral behavior *now*. Casuistry, by honoring both the multi-variegated complexity of moral conflicts as well as the need for normative principles, is essential to ethical inquiry, and Christian moral reasoning is the richer for it.

Faced with the ethical complexities that arise when trying to respond morally to intractable human conflict, the Christian, discipled under just war's tutelage, asks the essential question: What does *love* look like now, in *this* moment, and in *these* circumstances? Love, guiding every exertion in the life of the believer, authorizes the Christian to come to the aid of the threatened even as it requires proportionality, discrimination, compassion, and, when possible, mercy to those who threaten. As a casuistic moral framework, just war reasoning provides an accessible criteria by which we can guide our moral intuitions, imaginations, and decisions in specific cases in light of fundamental theological principles and moral norms. The just warrior, *while* warring justly, always attempts to apply the dictates of love, not to shirk them.

The Moral Necessity of the Bomb

How does a doctrine of military necessity, nested within a just war framework grounded in the imperative of neighbor love, countenance the killing of 80,000 souls in an atomic *pikadon*? An exhaustive defense of the bombing of Hiroshima is beyond the capacity of this chapter. There have been several excellent book-length treatments—and more coming—that do an admirable job.[14] Defense of the bombing has, by now, almost certainly become the minority view. The percentage who disapprove increases as the Second World War generation and their immediate descendants pass away. Against claims, like the one I'm making here, that the bomb was militarily and morally necessary to compel the Japanese surrender, critics have mounted multiple challenges. Richard Frank, the foremost historian of the Pacific War, identifies among these challenges a common foundation of three basic premises:

> First, that Japan's strategic position in the summer of 1945 was catastrophic. Second, that its leaders recognized their hopeless situation and were seeking to surrender. Finally, that access to decoded Japanese diplomatic communications armed American leaders with the knowledge that the Japanese knew they were defeated and were seeking to surrender.[15]

Thus, accuse the critics, American leaders, knowing the atomic bombs were not necessary to end the war, subjected innocent Japanese to nuclear devastation for other reasons, including: to justify the enormous expenditure of funds, to satisfy a perverse intellectual curiosity, or to intimidate the Soviets.[16]

One such critic, the Oxford philosopher Elizabeth Anscombe, famously gave a speech to her colleagues in 1956, attempting to persuade them to join her objection against Oxford's granting an honorary doctorate to President Truman. Happily, the measure failed (only three other faculty joined her dissent: her husband, her friend, and her friend's husband). In a pamphlet that followed, *Mr. Truman's Degree*, Anscombe lays out an almost withering—if at times overblown—argument against dropping the bomb. She admits that "given the conditions," the bombing "pretty certainly saved a huge number of lives." But everything hinges on what those "conditions" were and who brought them about. To her mind, the conditions that—appeared—to necessitate the atomic bomb was the unjustifiable Allied "fixation on unconditional surrender" and "the disregard of the fact that the Japanese were desirous of negotiating peace."[17]

Her first move therefore is to insist the bombing was not necessary. "It seems to be generally agreed," she writes, "that the Japanese were desperate enough to have accepted the [Potsdam] Declaration but for the loyalty to their Emperor."[18] With that in place, Anscombe lays bare her primary objection: the atomic attack intentionally killed innocent people as a means to an unnecessary end. This, she insists, is always murder, and any effort to justify can only result in the diminishing what distinguishes murder from morally permissible killing.

Let's turn first to the claim that the Japanese knew they were defeated and were willing to surrender. It's certainly true that by the summer of 1945, Japan was defeated. It's also true, as intercepted and decoded Japanese military and diplomatic radio communications made clear at the time, that by at least as early as the fall of 1944 the Japanese military and political leadership knew this. From the beginning of 1943 onward, Japan had sustained almost nothing but an unbroken series of reversals in the face of the American counterattack. The decisive catastrophes the Empire endured in the summer and fall of 1944, culminating in their devastating defeats at the Battles of the Philippine Sea—where they lost the air—and Leyte Gulf—where they lost the sea—set in motion the ultimate fate facing Japan. Their loss on the water off Leyte opened up the island's beachheads to Allied invasion. The subsequent inability of Japanese land forces to dislodge the invaders in turn signaled the inevitable loss of the Philippines in its entirety. This was a blow from which the Japanese could not—and did not—expect to recover. Japan now knew herself to be irreversibly cut off from the territories she occupied in Southeast Asia, resulting in the loss of access to materials—such as oil and food—essential to continuing to prosecute its war with any reasonable hope of success.

Nevertheless, neither Japanese defeat nor Japanese recognition of their defeat corresponded to a Japanese willingness to surrender. This shouldn't be surprising. In over 2,000 years of Japanese history, no Japanese ruling authority had every surrendered to a foreign power. Closer to home, over the course of the Pacific War, not even a Japanese *unit* had ever surrendered. A wealth of evidence uncovering the disposition of the Japanese leadership in the leadup to Hiroshima, some of the most important of which has only been available since the mid-1990s—including records of intercepted radio intelligence, minutes from Japanese war cabinet meetings, and postwar interrogations—puts to rest any serious suggestion that prior to August 6th, 1945, Japanese leadership, including the emperor, was anywhere close to permitting them to turn from knowledge of defeat to surrender.[19]

Undeterred, those who insist on arguing that American policymakers knew that Japan was near surrender when they chose to bomb Hiroshima point to a series of intercepted diplomatic cables exchanged from July 11 to August 3 between Japanese Foreign Minister Shigenori Togo in Tokyo and Japanese Ambassador Naotake Sato in Moscow as proof of the diplomatic effort by key Japanese leadership to secure the Soviet Union as a mediator to negotiate an end to the war.

But gesturing to these cables to argue Japanese openness to an early peace is a fool's errand. Taken as a whole, the diplomatic exchange demonstrates the Japanese government's lack of earnest desire to negotiate terms. A serious diplomatic initiative would require two things: (1) concessions by the Japanese that would enlist the Soviets as mediators; and (2) Japanese terms to end the war. But to Sato's evident frustration, "Japan never completed either of these two fundamental steps." Over and over again, Sato rebukes his leadership's unwillingness to accept reality. "If the Japanese empire is really faced with the necessity of terminating the war," he insisted, "we must first of all make up our minds to terminate the war." The Japanese government's continued equivocation only proved Sato's fear that they lacked any real intent to do so.[20]

The Sato-Togo exchange is particularly helpful in dispelling a particularly resilient myth. As Anscombe charged, the most persistent condemnation of American diplomacy in the summer of 1945 is that policymakers understood that a promise to retain the Emperor would have likely resulted in a prompt surrender. We needn't wonder about this. In late July, Ambassador Sato advised Foreign Minister Togo that the best terms Japan could hope to secure were unconditional surrender, modified only to the extent that the Imperial institution could be retained. Presented by his own ambassador with this offer, Togo shot back a reply that this was wholly unacceptable. "Given this," Frank concludes, "there is no rational prospect that such an offer would have won support from any of the other five members of the Supreme Council for the Direction of the War."[21]

This is a critical insight. The "Big Six" was a select body that comprised the ultimate authority within the legal government of Japan. It included the prime minister, foreign minister, army minister, navy minister, chief of the Imperial Army General Staff, and chief of the Imperial Navy General Staff. The other two individuals who exercised real power were the emperor and his principal advisor, the Lord Keeper of the Privy Seal. The Big Six were gridlocked on the question of terminating the war by a procedural rule requiring complete unanimity among them in order to make decisions. Effectively, then, each individual member carried a veto. With four serving officers of either the Army or Navy among the six, the militarists held indisputable power. They wielded it.

While Japanese leadership knew relatively early that they could never conquer the United States in a full-contact brawl, the militarists pressed the gamble that they would not have to. Against the claim that Japanese recognition of their hopeless situation obviated the need for the bomb, Frank asserts that "the harsh reality is that the key Japanese leaders in the summer of 1945 *did not* view their situation as hopeless."[22] This was neither myopia nor stupor, it was strategic. Japanese leadership, argues Frank, weren't "simply staggering on with the war in a fanatical trance, oblivious of their actual plight." On the contrary, the Japanese simply pinned their hopes on something other than victory: a "coherent and thoughtfully conceived military and political strategy called *Ketsu Go* (Decisive Operation)."[23] Understanding *Ketsu Go* and the investment of Japanese leaders in this strategy is essential in understanding the military necessity of the bombing.

The core premise of *Ketsu Go* was that American morale was brittle. Sustained by their faith in their own racial superiority, and fueled by the belief that Americans lacked the spiritual stamina of the Japanese, Japan's leadership fortified themselves by the delusion that a prolonged war with increasing casualties would see America's modest morale crumble. With the American will to see the war through sapped, American political leaders would negotiate an end to the conflict on terms favorable to the Japanese.[24] While by 1945, this Japanese fever-dream ought to have been thoroughly dashed by the reality of American resolve, they held to the belief that if they could bloody the United States enough in the war's closing battles, they could still stand down with negotiated terms that would not bring dishonor.

In January 1945, Japanese prescience kept the possibility of a final fight alive for them. They figured out where the American invasion of the Japanese homelands would begin. The discovery wasn't a matter of espionage, code breaking, or clairvoyance. They simply drew on what they knew about American character, operational technique, and goals and then coupled this knowledge with the expectation that by the time of invasion the most advanced US air bases would be on Iwo Jima and Okinawa. As Frank lays it out:

Although the Japanese much feared the possibility that their adversaries might try to force surrender with a campaign of blockade and bombardment, they believed Americans lacked the patience for [this]. Therefore, they would invade… American superiority in combat power during the Pacific War rested upon overwhelming air and sea power, not ground forces. It followed that U.S. plans to invade Japan must encompass the ability to bolster their ground units with masses of planes and ships….Okinawa provided the capacity to support several thousand aircraft. Iwo Jima did not. Thus, an arc representing American fighter plane range from Okinawa foretold the likely American landing areas. Within that arc fell Kyushu and parts of Shikoku. Compared to Shikoku, the southern ranges of Kyushu around Miyazaki, Shibushi Bay and the Satsuma Peninsula formed the most obvious targets with plentiful airfield sites and naval bases from which the Americans could mount an invasion of the Kanto (Tokyo) plain.[25]

What followed was a flood of refortification. On New Year's Day 1945, there were only 12 field divisions in all of Japan. By the time Imperial Headquarters completed their massive program of homeland reinforcement, there would be 60 divisions (36 field and counterattack, 22 coastal combat and two armored divisions) and 34 brigades (27 infantry and seven tank).[26]

By July and August, intercepted Japanese communications had revealed to American leaders the ambush awaiting them on Kyushu alone, the first phase of the invasion plan. US planning had been predicated on the assumption that the 680,000 American fighting men slated for the invasion of Kyushu would face no more than 350,000 Japanese. But the decrypted communications unmasked the massive Japanese buildup. From a single field division in January 1945, the defenses on Kyushu now consisted of 14 field divisions, three tank brigades, and eight independent mixed infantry brigades. The aggregate forces comprised some 900,000.[27]

Augmenting the numerical depth, *Ketsu Go* encompassed a comprehensive devotion to *tokko*, a euphemism for suicide attack. These would include not only by this time routine air and sea suicide attacks but also land based now as well. Moreover, while originally expecting to face only 2,500–3,000 aircraft, by midsummer, American intelligence confirmed the Japanese could field nearly 10,000 aircraft, about half earmarked for *kamikaze* attacks, which would pummel the invasion convoys.

Ghastlier still, *Ketsu Go's* most singular feature was the massive incorporation of the civilian population into the defense scheme. In March 1945, Public Law Number 30 mobilized every able-bodied male age 15–60 and every female 17–40.[28] This inducted about a quarter of Japan's total population. Frank's assessment is worth quoting at length:

> The significance of these plans cannot be exaggerated. This mobilization aspired to create from the mass of the population a huge pool of untrained men and women who would be married to tactical units, where they would perform direct combat support and ultimately combat jobs. This would literally add tens of millions to the strength of the ground combat units, albeit of little formal combat power for lack of training and equipment. It would also guarantee huge civilian casualties and make a reality the disturbing American nightmare of a "fanatically hostile population." By mustering millions of erstwhile civilians into the area swept by bombs, artillery, and small arms fire, Japan's military masters willfully consigned hundreds of thousands of their countrymen to their deaths. Moreover, by deliberately obliterating any distinction between combatants and noncombatants, they would compel American soldiers and marines to treat virtually all Japanese as combatants, or fail to do so at their peril.[29]

There was no reason for American military or political leaders to disbelieve Japan's willingness to commit national suicide. The 1945 Battles of Iwo Jima and Okinawa were horrific representations of this appalling determination. As if made to purpose, Iwo Jima—or Sulphur Island—is eight square miles of spewed earth some 650 miles south of Tokyo. As a physical thing, it was decried for its ugliness. William Manchester in *Goodbye, Darkness,* his memoir of the Pacific war, denounces Iwo as "an ugly, smelly glob of cold lava squatting in a surly ocean."

Its porkchop-shaped landscape—already conjuring images of butchered meat—is an unrelieved gray, gray-green, brown, and black, the hues of camouflage. It's as if the reeking island was gestated and purpose-bred for war. Just so, from February 19 to March 26, 1945, a total of 6,821 Americans and approximately 20,000 Japanese died in the fight. Twenty thousand more Americans would be wounded. Not so the Japanese. Estimates vary, but only about 216 of the 21,000-strong Japanese force survived the fight. It was unintentional. They were simply too wounded to continue fighting or to kill themselves. Iwo was the only island battle of the Pacific War in which US casualties outnumbered Japanese. Following Iwo Jima, and as if trying to top it, the Battle of Okinawa proved the bloodiest fight of the Pacific War. The imperial war machine threw upward of 225,000 souls into its suicidal scheme, including around 150,000 civilians and 77,000 warriors. The civilian count includes some 40,000 Okinawan civilians the Japanese army pressed into combat.

The American experience of these and earlier island assaults combined with the revelations of the buildup on Kyushu, provoked American leaders into an agonizing review of whether the invasion of the Japanese homeland was even still viable. As General MacArthur's intelligence officer quipped, a ratio of one American for every Japanese defender "is not the recipe for victory."

This rather long disquisition on the Japanese defensive plans and warrior ethos is worthwhile because it helps set up why the demand for unconditional surrender was necessary. It tells us something about the character of Japanese militarism and the lengths Japanese militarists were willing to go—and the horrors into which they were willing to take their own people and subject others—in order to save face or to continue to pursue their imperial delusions. Having refused, in their intractable arrogance, to stand down when it was clear they had lost the fight, the Japanese essentially proceeded to demand that we collude with them in the perverse idea that the fight should continue. It's against this attitude that the notion of unconditional surrender being "not simply a slogan about victory, but a policy about peace" comes to light. There are at least two ways this plays out.

First, for many of the senior American military and political leadership, the First World War was a relatively recent memory. Some of them might even have recalled the Commander in Chief of the American Expeditionary Forces General John Pershing's strongly worded warning, in the midst of the armistice deliberations, that the Allies should refuse to grant Germany any terms whatsoever and should, instead, press their attack against the Kaiser without quarter. Pershing feared that Germany didn't "know that she was licked."[30] The general believed that a beaten enemy is the more easily compelled toward a durable peace. A decisive victory, having taken the fight out of the enemy, allows for a more realistic hope than a weak armistice that the matter has truly been settled and that the contest will not have to play out again. The simple fact that someone is *not* shooting at you does not mean that he does not *want* to or that he will not if given half a chance. As it turns out, history was on Pershing's side. The Treaty of Versailles left Germany neither pacified nor conciliated nor weakened beyond recovery. One or the other might have been sufficient to prevent the next great war.

What this suggests is that we should view decisiveness as an implication of the just war tradition. While the "right intention" criterion of the *jus ad bellum* framework is peace, it's not just any kind of peace that is in view. The end of a just war ought to be a peace that sufficiently allows for the establishment or reestablishment of a political community that is at least roughly characterized by conciliation, order, and justice. It is toward this kind of outcome that the force in war is to be properly proportioned. It is a mistake to conceive of proportionality as having economy of effort or restraint as its basic imperative. It is true that combatants are required to employ only as much force as is necessary to achieve legitimate strategic and tactical objectives and as is proportionate to the importance of those objectives. The just warrior must be neither gratuitous nor excessive. But in place of restraint, the basic imperative of proportionality is the deployment of that amount of force sufficient for a decisive victory aimed at a durable peace.[31] Clearly, the enemy has a say in how much force will be required to bring this about.

That notion brings us to the second means by which the demand for unconditional surrender was a policy for peace: it provided the legal authority to execute the far-ranging political changes needed to renovate the internal structure of the Axis nations, ensuring that they would never again pose a threat to their neighbors.[32] More than just an imperial army needed to be defeated in the Asia-Pacific in 1945. An idea needed to die. A mythology dating from 1853 in response to foreign intrusion into Japanese life gave rise to a malignant national essence which strove for "the submission—indeed, the subsumption—of the individual to the imperial will and the state." The result, writes John Lewis, was

> the inculcation of theological militarism into Japanese culture, a synthesis of selfless devotion to the nation, the race, and the emperor with long-standing military ideals in a rigid structure.... With this political, social, and educational system in place, and under military leadership that was motivated to attain a place of dominance in Asia, the Japanese set out to create an empire, "as befitted their destiny as a superior race."[33]

What followed was, in essence, a holy war. As such, gods would have to be killed. "To roll back and end Japan's drive for empire," Lewis insists, "would require more than military action—serious changes would have to be made *inside* Japan. Before such changes could be made, however, Japan would have to be thoroughly defeated militarily"[34] and made thoroughly convinced of that defeat. The requirement of unconditional surrender signaled a commitment to "the total and permanent destruction of Japan's will and capacity to fight."[35]

This was the military—and moral—vision for unconditional surrender. Entirely different from an armistice agreement reached by negotiations, unconditional surrender began with a *demand*.[36] When coupled with the atomic attacks and American intransigence, the issue of surrender became an either-or proposition for the Japanese. "The complete loss of hope was central to Japan's decision to surrender," Lewis says. "As long as its leaders saw even a slim chance of preserving their system, they would grasp at that chance." The *pikadon* evaporated that hope. There would be no great battle, no banzai charge, no preservation of the military system. It was a mercy. "Sixty years of indoctrination had created a cultural straitjacket that could be removed by nothing less than overwhelming power and intransigence."[37]

The possibility of such a transformation was nowhere on the visible horizon prior to the *Enola Gay* lifting off. Even after bombs detonated over Hiroshima, then Nagasaki, and even after the Soviet declaration of war cut off any possibility for a brokered peace sympathetic to Japanese terms, the Big Six as a body never initiated a move toward peace. Paralyzed by a militarist cabal that still sought to win concessions that the Allies would never offer: the right to repatriation of its own armed forces and no demobilization; sole jurisdiction over so-called war crimes trials; and no occupation of the homeland. At all costs, the cabal sought to preserve the Japanese "national essence" and retain its militaristic theocracy. Only Emperor Hirohito's shocking intervention overrode the deadlock and denied the fanatics their final paroxysm of violence. Whatever his previous complicity, in the wake of those horrible bombs, the Emperor finally called upon the powers of his station to exorcise from the Japanese their malignant resolve to perish as a people rather than surrender.

But whatever one contends about the morality of demanding unconditional surrender, we still have the problem of the innocent Japanese dead. About those dead, Anscombe insists that "for men to choose to kill the innocent as a means to their ends is always murder." With the bombing of Hiroshima in view she concludes, "it was certainly decided to kill the innocent as a means to an end."[38] Hiroshima, therefore, was murder. And murder must never be done.

But was "choosing to kill the innocent as a means to an end" really what happened over Hiroshima? Or, *if* it was, was it *simply* that? For instance, it is important to recognize that throughout 1945, across the Asia-Pacific, large masses of innocent people were dying and being killed daily. It is morally important to recognize that every one of them owed their plight, ultimately or directly, first to the originating Japanese aggression that started the war in the first place and, second, to the present Japanese refusal to stand down when everything was already lost. Moreover, because of the circumstances engineered by Japanese choices, these groups of innocents were in conflict with one another, in the sense that they had competing interests. Goods done for one group often meant harms were done, or allowed to be done, to other groups. Who were these clusters of innocents? And how do their competing interests affect our assessment of the necessity of the atomic bombing?

Most on President Truman's mind were the hundreds of thousands of American warfighters preparing for one last brawl. The vast majority of them were either post-Pearl Harbor volunteers or conscripts. By June 1945, American young men, despite the defeat of Nazi Germany being already in the books, enjoyed no peace dividend. The United States was already several months into the steep increase in draft calls implemented under President Franklin Roosevelt in order to produce a 100,000-men-per-month "replacement stream" in preparation to meet casualty projections for the upcoming invasion of Japan. By D-Day, some 600,000 new recruits would need to be trained, equipped, and deployed.[39]

Of course, Anscombe wouldn't countenance American warfighters—or any warfighter—being included in a tally of innocent groups. She wouldn't mean that every American sailor, soldier, airman, or marine was somehow personally culpable for the fighting in general or for particular acts of unjust fighting. By "innocent," she was simply referencing those who are protected under the discrimination requirement, those "who are not fighting and not engaged in supplying those who are with the means of fighting." Innocent doesn't mean blameless here, it means "not harming."[40] Those who are non-harming in war are noncombatants and ought not to be touched. Anscombe gives no quarter to conscripts on this, whatever side they may be on. Conscripts, having accepted conscription and joined the fight, are now "harmers" and have waived their right not to be harmed in turn.

This is largely right, but especially—though not exclusively—with conscripts in mind I endorse a slight loosening of the doctrine of discrimination in particular situations. I do not say unraveling. As already mentioned in the introduction to this volume and various places throughout—particular in the chapters by Hastey and Braun—the doctrine of military necessity includes the promotion of stewardship. In the pursuit of military objectives, a commander—wherever they fall along the chain of command—bears a moral obligation for the responsible expenditure of his nation's resources—materiel, treasure, and especially human. There are rough, context-dependent limits to the amount of risk to which it is appropriate for a commander to subject a warfighter even if imposing those limits simultaneously increases risk to civilians, meaning the adversary's civilian population or those populations sympathetic to the adversary. Any such loosening of discrimination would require, first, that there are no equally efficacious alternatives both to reducing the risk to warfighters at the expense of civilians and to achieving the military objective or war aim and, second, that the rebalancing of risks is proportionate—in terms of both the decreased risk to warfighters and the increased risk to civilians as well as between the increased risk to civilians and the goods that may be enjoyed and the harms avoided by achieving the military objective. Stewardship is an essential command responsibility, for while a commander must be willing to spend the lives of his men, he must never be willing to waste them.

But what then of the Japanese conscript? Do I include them in my tally of innocent groups? Yes, but with qualifications. Stewardship cuts both ways. Choices have consequences. A sovereign's choices rain down consequences equally upon those citizens who agree with those choices as well as those who disagree—and, most tragically, on those as yet too young to do either. It is morally significant that on one side of a conflict, there may be the obvious aggressor nation and, on the other, a nation which is simply responding to the aggression in order to protect the innocent, to requite injustices, and to punish evil. Because of this, there seems to be basic distinctions in responsibility—which is not necessarily the same thing as guilt—that can be identified between either an American conscript or a post-Pearl Harbor volunteer and a Japanese soldier fighting zealously for empire—and even between the American conscript and his Japanese counterpart. It may be true that the average conscript—whether American or Japanese—was, in the first place, not a warfighter at all but rather a teacher, or a plumber, baker, farmer, clerk, husband, father, son, or brother and is now, in the second place, only a reluctant, if obedient, soldier. It's possible, even likely, that both of them would equally rather have been anywhere else than fighting in the Asia-Pacific. Nevertheless, one of them is there because his country was unjustifiably under attack. The other because his country was unjustifiably attacking it. Surely, that must sometimes issue in giving certain preferences for the former over the latter. President Truman, with his own young men in view—those brothers, sons, fathers, husbands, clerks, bakers, and plumbers—and with casualty projections mounting, rightly committed himself to any course of action—including putting at risk Japanese plumbers, and bakers, and clerks—that might allow him to avoid throwing his boys into one bloody Okinawa after another from one end of Japan to the other.[41]

What of the other clusters of innocents? While military concerns naturally form the core of strategic decision-making, military necessity and humanitarian interests ought both to inform leadership decisions. Hitlerism's staggering assault on humanity is well-known and well-documented. But unlike the millions lost to Nazi rapaciousness, beyond, perhaps, a vague awareness of the infamous massacre at Nanking or the subjugation of Korean women, most Americans probably know little about the unfathomable human toll of Japan's bestial aggression on populations across the Pacific. They should know, for the details matter. Lewis gives us a clearer understanding of the meaning of Japanese occupation in his summary description of the "Rape of Nanking."

> [The atrocity] may have killed 300,000 Chinese…Thousands of women were gang-raped and forced into military prostitution. Thousands of Chinese civilians were herded and machine-gunned, used for bayonet practice, buried alive, doused with gasoline and burned, or decapitated with swords before smiling Japanese troops. The Japanese media covered the killing contests; the *Japan Advertiser* ran pictures of two officers who competed to see who would be the first to kill one hundred men with a sword.[42]

The Second World War in the Pacific was launched, and prosecuted, by a nation whose highest ideals had been made, in Lewis' words, "violently hostile to human life." But whatever contemporary Americans know now about conditions in the war-torn Pacific, at the time key Allied decision-makers would have been very well informed indeed. Various entities, including the US ambassador in Chungking, British and American military commanders in theater, and the Office of Strategic Service teams in China, Thailand, and Indochina, kept Washington informed about the suffering of Asia.[43] Werner Gruhl summarizes the macabre accounting:

> By August 1945 the Allies in the Far East and Pacific had paid a price for the long years of resistance to aggression that is incomprehensible to most of us today. In this vast geographic crucible, some 20 million innocent Allied civilian lives, of whom at least three million were children, were snuffed out by the war. Another estimated 85 million civilians or more suffered forced labor

and refugee ordeals, malnutrition and disease, wounding, maiming, rape, and torture; internment hell, war orphan and widow anguish, and Japanese supported opium addiction.[44]

Japanese noncombatant deaths may have reached, at the upper limit, 1.2 million. A rough calculation, then, tells us that for every Japanese noncombatant death, some 17 or 18 noncombatants from other Asian nations died—and 12 of them would be Chinese.

Japan's military controlled the destiny of the Far East and could have brought an early end to the suffering with a word. But because of its continued refusal to ratify its own defeat, by August 1945, each added week of war placed another 100,000 Chinese and other allied civilian deaths on the butcher's bill. Every day that the war ticked on, up to another 14,000 civilians died.[45]

The third cluster of innocents were Japanese civilians themselves. Japan, of course, while they doled out staggering amounts of death, sustained death as well. When the Soviets overran Manchuria and other areas on the Asian continent, they captured around 2.7 million Japanese, about two-thirds of whom were civilians. Between 340,000 and 370,000 disappeared in Soviet captivity forever. In the homeland itself, Gruhl estimates that by the end of the summer of 1945, upward of 50,000 Japanese died each week. As a cluster of innocent people, the Japanese civilian population was also, if abstractly, in conflict with itself. What I mean is this: there were only so many scenarios by which the war was going to end. Different amounts of Japanese civilians were going to die in each scenario. It is instructive to compare those numbers. For instance, under the blockade and bombardment campaign, Japanese were presently being killed by the relentless American bombing runs. But as the casualty estimates for the proposed land invasion continued to mount, interest began to increase for simply tightening the blockade and keeping Japan under siege until she surrendered. The cruelty of this strategy would have been mindboggling. It would have resulted in millions of starvation deaths. Had American planners stuck with the invasion, the numbers of Japanese civilian deaths would likely have been less than in a blockade, but they would certainly have far exceeded the deaths due to the atomic bombs, especially given the universal conscription orders and the shocking willingness to die rather than give in. In the macabre world the Japanese made, one can conjecture across time and space the existence of various clusters of innocent Japanese civilians whose interests were in competition with one another. It was the Japanese leadership who determined that some cluster of its innocent civilians or another would die before the war ended. The only thing the Americans could do was to limit that death as much as their available options made possible.

The point is this. For every day that the war continued, thousands upon thousands of innocent civilians died throughout the Asia-Pacific. Every day that Japanese leadership continued to gamble on a decisive final fight, their intransigence was measured in the heaping mounds of the innocent dead. Anscombe's insistence on the inviolability of innocent lives carries a strong claim. But it collides with history. The tragic reality is that there were clusters of innocent lives scattered across the Pacific and not all of them could live. Every decision, every action or inaction, would doom one or another of them.

The bombs offered the Allies the best possible means of bringing the war to a speedier end and in a manner decisive enough to give the best possible hope for a durable peace. Every alternative to the bombs would have resulted in a grotesquely greater death toll among both Allied and Japanese combatants and civilians. In light of everything, there was cause to prefer some clusters of innocent lives over others. Anscombe's insistence that one must not "choose to kill the innocent" is important, it is also very imprecise. One cannot say—though at several points Anscombe appears to—that American leadership chose the killing of innocents in Hiroshima in quite the same way as the Nazis chose to kill the innocent at Auschwitz. Taking a more contemporary case, Truman did not desire the deaths of the people of Hiroshima in the same way that William Calley desired the deaths of innocents in My Lai. There is more that can and should be said about different kinds of intention and how they play out in the close of the Asia-Pacific War, but this is not the place for it.[46] As a part of that discussion, much more ought to be said about why Hiroshima was the target city. Again, space prohibits that discussion here save for but a hint of it. As I have written elsewhere,[47] if the intention was truly to kill innocent people as a means to an end, there would have been other, more populous, choices. Hiroshima, with its structures largely untouched due to never having been previously attacked, was the best option for demonstrating, without ambiguity, the power of the terrible new bomb.

In any case, a shortened war was a boon to the lives of the innocent. For American leadership and beleaguered people throughout the Pacific, locked in a contest with Japanese militarists and popular sentiment that preferred national suicide to defeat, the promise of the bomb must have felt like deliverance. Truman's decision to hit Hiroshima—and later Nagasaki—ultimately killed, perhaps, 200,000 people, the majority of them Japanese civilians (though about 20,000 military personnel were also killed). While tragic, those Japanese civilians had no greater claim to not be harmed than the Chinese and other Asian civilians dying under Japanese occupation throughout the Pacific, or the Japanese noncombatants suffering in Soviet captivity, or the Japanese noncombatants—and Allied prisoners of war—who would have died on the Japanese home islands as the war dragged on, or, as I have argued, no infinitely greater right not to be harmed than the young American conscripts dragged across the ocean to finish a war that ought already to have been concluded. Moreover, as I have stressed about the Japanese civilians themselves, there was cause—if forced to choose—to prefer the lives of other civilians over theirs. This, too, is tragic. But it is a tragedy of Japanese engineering.

There can hardly be a happy ending for something as grim as the Asia-Pacific War. But history has shown that Japan would have better days ahead. Those good days are a part of the bounty of Hiroshima. As the historian John Dower put it:

> Because the defeat was so shattering, the surrender so unconditional, the disgrace of the militarists so complete, the misery the 'holy war' had brought home so personal, starting over involved not merely reconstructing buildings but also rethinking what it meant to speak of a good life and good society.[48]

The history seems clear: the atomic bombing of Hiroshima offered the most moral—that's to say, the most proportionate and discriminate—means possible to obtain the unconditional surrender of the Japanese in the quickest possible time. Military necessity and humanitarian concern stood in harmony. In the context of the Asia-Pacific War in the summer of 1945, *because* the bombs were necessary, they were moral.

Notes

1 "The Bombing of Hiroshima," The Atomic Archive, accessed March 2, 2023, https://www.atomicarchive.com/history/atomic-bombing/hiroshima/index.html.
2 From testimony by Setsuko Thurlow, a *hibakusha*—survivor—of the Hiroshima bombing. https://www.redcross.ie/blog/2014/12/a-13-year-olds-hiroshima-experience/.
3 Of course, between Hiroshima and Japanese surrender was both the second atomic attack—on Nagasaki—and the Soviet declaration of war against Japan. Two comments. First, my reference to "Hiroshima" serves, largely, as shorthand for both atomic attacks. Second, while Soviet entry played a crucial role, I contend that it was the atomic bomb that played the more significant role in getting Japan to surrender how and, most importantly, *when* it did. This point won't be directly argued in force here. It has been argued so by others (Richard Frank, William Miscamble, Sadao Asada, and Robert Newman among them). I argue it so myself in my forthcoming book, *Moral Horror: A Just War Defense of the Bombing of Hiroshima* (2024).
4 Steven P. Lee, *Ethics and War: An Introduction*, Illustrated edition (Cambridge; New York: Cambridge University Press, 2012), 216.
5 Reinhold Niebuhr, "Our Responsibilities in 1942," *Christianity and Crisis* 1, no. 24 (1942): 1–2.
6 For an extended discussion of the Niebuhrian paradox between love and justice, see the second chapter of my book, "The Problem of Paradox" in Marc LiVecche, *The Good Kill: Just War and Moral Injury* (New York, NY: Oxford University Press, 2021).
7 "The Bombing of Germany," in Reinhold Niebuhr, ed., *Love and Justice: Selections from the Shorter Writings of Reinhold Niebuhr* (Louisville, KY: Westminster/John Knox Press, 1992), 222.
8 Daniel M. Bell, *Just War as Christian Discipleship Recentering the Tradition in the Church Rather than the State* (Grand Rapids: Brazos Press, 2009), 91.
9 Michael Walzer, *Just and Unjust Wars: A Moral Argument with Historical Illustrations* (New York: Basic Books, 1977), 251–268.
10 For an excellent summary of Walzer on supreme emergency see: J. Daryl Charles and Timothy J. Demy, *War, Peace, and Christianity: Questions and Answers from a Just-War Perspective* (Wheaton, IL: Crossway, 2010), 220–224.
11 Michael Walzer, *Arguing About War* (New Haven, CT: Yale University Press, 2004), 34.
12 Walzer, *Arguing About War*, 34–35.
13 In addition to the Frenchs' chapter in this volume, see, for example, LiVecche, *The Good Kill*; LiVecche, "Kevlar for the Soul: The Morality of Force Protection," *Providence: A Journal of Christianity & American Foreign Policy* Fall 2015 (January 11, 2016), https://providencemag.com/2016/01/kevlar-for-the-soul-morality-force-protection/.
14 See, for example: Richard Frank, *Downfall: The End of the Imperial Japanese Empire* (New York: Penguin Books, 1999), chap. 16 "Hiroshima"; Werner Gruhl, *Imperial Japan's World War Two: 1931–1945* (New Brunswick, NJ: Routledge, 2006); and Tom Lewis, *Atomic Salvation: How the A-Bomb Saved the Lives of 32 Million People* (Casemate, 2020). For two more theological treatments see: Francis X. Winters, *Remembering Hiroshima: Was It Just?* (Farnham, England; Burlington, VT: Routledge, 2009); and especially Wilson D. Miscamble C.S.C, *The Most Controversial Decision: Truman, the Atomic Bombs, and the Defeat of Japan* (New York: Cambridge University Press, 2011).

My own, *Moral Horror: A Just War Defense of the Bombing of Hiroshima,* extends the argument of this essay by arguing against the proposition, present in the theological accounts, that the atomic bombings were "lesser evils." I do so partly by demonstrating how each of the criteria in the just war framework argue in favor of the bombings. I conclude that doing anything else in the summer of 1945 would have been morally unjustifiable.

15 Richard Frank, "Ketsu Gō: Japanese Political and Military Strategy in 1945," Tsuyoshi Hasegawa, ed., *The End of the Pacific War: Reappraisals,* 1st ed. (Stanford, CA: Stanford University Press, 2021), 65.

16 Frank's summary of themes from the postwar critical literature is drawn from J. Samuel Walker, "The Decision to Use the Bomb: A Historiography Update," *Diplomatic History* 14 (Winter 1990): 97–114, and Barton J. Bernstein, "The Struggle Over History," in Philip Nobile, ed., *Judgment at the Smithsonian* (Notre Dame: Marlowe, 1995), 127–256, esp. 162–167, 173, 178, 195–198.

17 G. E. M. Anscombe, "Mr. Truman's Degree," in *Ethics, Religion and Politics* (Oxford: Blackwell, 1981), 65.

18 Anscombe, *Ethics, Religion and Politics,* 64.

19 While the intercepts do confirm that a number of Japanese diplomats in Europe attempted to open negotiations to end the war, none of them were acting with any actual governmental authority. For rebuttal against claims that early surrender was on the table see: Frank, *Downfall: The End of the Imperial Japanese Empire,* esp. pages 214–139; and Miscamble, *The Most Controversial Decision,* esp. pages 94–111. For a fascinating, if brief, summary of the importance of the Magic program, see: Richard Frank, "Why Truman Dropped the Bomb," *Washington Examiner,* August 8, 2005, available at: https://www.washingtonexaminer.com/weekly-standard/why-truman-dropped-the-bomb.

20 A tidy overview of the Sato-Togo diplomatic exchange can be found at: "'Pretty Little Phrases': Japanese Diplomacy in 1945," The National WWII Museum | New Orleans, August 14, 2020, https://www.nationalww2museum.org/war/articles/japanese-diplomacy-1945.

21 Frank, *Downfall: The End of the Imperial Japanese Empire,* 239.

22 Frank, "Ketsu Gō: Japanese Political and Military Strategy in 1945," 65.

23 Frank, 65–66. A noteworthy—and under-reported—fact is that *Ketsu Go* was commanded by Japan's 2nd General Army, headquartered at Hiroshima.

24 Frank, 68.

25 Ibid., 72.

26 Richard Frank, "Kemper Lecture 2001" (America's National Churchill Museum), available at https://www.nationalchurchillmuseum.org/kemper-lecture-frank.html.

27 Frank, "Ketsu Gō: Japanese Political and Military Strategy in 1945," 74.

28 Frank, *Downfall: The End of the Imperial Japanese Empire,* 85–86.

29 Frank, "Ketsu Gō: Japanese Political and Military Strategy in 1945," 77.

30 Donald Smythe, *Pershing: General of the Armies,* 1st Pbk. Ed edition (Bloomington: Indiana University Press, 2007), 232.

31 For more on this, see: Marc LiVecche, "Grim Virtue: Decisiveness as an Implication of the Just War Tradition," in Timothy S. Mallard and Nathan H. White, eds., *A Persistent Fire: The Strategic Ethical Impact of World War I on the Global Profession of Arms* (Washington, D.C.: National Defense University Press, 2020), 21–46.

32 Frank, "Ketsu Gō: Japanese Political and Military Strategy in 1945," 66.

33 John David Lewis, *Nothing Less than Victory: Decisive Wars and the Lessons of History* (Princeton, NJ: Princeton University Press, 2010), 242.

34 Lewis, *Nothing Less than Victory,* 246.

35 Ibid., 248.

36 Ibid., 264.

37 Ibid., 262.

38 Anscombe, "Mr. Truman's Degree," 64.

39 D. M. Giangreco, *Hell to Pay: Operation DOWNFALL and the Invasion of Japan 1945–1947* (Annapolis: Naval Institute Press, 2020), x.

40 Anscombe, "Mr. Truman's Degree," 67.

41 For a thorough accounting of casualty estimates, see: D. M. Giangreco, ""A Score of Bloody Okinawas and Iwo Jimas": President Truman and Casualty Estimates for the Invasion of Japan," in Robert James Maddox, ed., *Hiroshima in History: The Myths of Revisionism* (Columbia and London: University of Missouri Press, 2007), 76–115.

42 Lewis, *Nothing Less than Victory*, 245.

43 Gruhl, *Imperial Japan's World War Two*, 211.

44 Ibid., 141.

45 Ibid., 145.

46 Among the best discussions of different kinds of "intent" is "The Principle of Double Effect: Can it Survive Combat?", the third chapter in: Nigel Biggar, *In Defence of War* (Oxford: Oxford University Press, 2013). One could also do much worse than the series of exchanges between Biggar and Christopher Tollefson beginning with Tollefson's "In Defense of the Innocent," then Biggar's response, "In Defence of Killing the Innocent, Deliberately But Not Intentionally," and finally, "Intention, Choice, and the Right to Life: A Response to Nigel Biggar. All three can be found at publicdiscourse.com

47 LiVecche, "Thinking About the Unthinkable," *Providence: A Journal of Christianity & American Foreign Policy.* https://providencemag.com/2015/08/thinking-unthinkable-2/.

48 John Dower, *Embracing Defeat: Japan in the Wake of World War II* (New York: W.W. Norton, 1999), 25.

13 Military Necessity

The Road Ahead

Eric Patterson

This book originated, in part, from the disjuncture between my own military training in the U.S. Air Force and reading the law of armed conflict, in contrast to most just war writing by scholars, especially in philosophy and theology, over the past half-century. Military officers are trained on *jus in bello* principles that include *military necessity*, as well as *proportionality* and *discrimination*. This is not just the formal training of those officers in the United States. It is also the training of the military officers in America's allies as well as the training held in foreign countries by agencies such as the International Committee of the Red Cross (ICRC). In short, our legal scholars and military personnel are trained on these *jus in bello* criteria, but most just war scholars in the social sciences and the humanities, myself included, have left *military necessity* out of the just war criteria.

As discussed in Chapter 1, if military necessity had the rare mention in a just war book in the past 60 years, it was either derided as Machiavellian *realpolitik*, or it was somehow mashed into the *jus ad bellum* criteria in terms of strategic necessity, i.e., defending against the potential extinction of a national group. The problem created by this gap in the literature has been exacerbated by some of the academic just war theorists, typically resting comfortably in philosophy departments, who have used just about every means available to them to put draconian restraints on warfighters in the field, contributing to an environment that severely constrains our men and women with restrictive rules of engagement when fighting enemies.

This volume calls for scholars to return *military necessity* to its rightful place in the *jus in bello* criteria as an important means of advancing just war as moral statecraft. Fortunately, common sense and law school textbooks on the law of armed conflict have kept *military necessity* alive, and for good reason. But it has done so in ways, as the preceding chapters have made clear, that are ultimately inadequate. In this concluding chapter, I would like to summarize the work that has been done here to help us better understand what military necessity is and why it is important for battlefield operations. With that in hand, we will turn our attention to areas where *military necessity* is of particular importance and to identify open avenues for additional study and application.

DOI: 10.4324/9781003390398-13

Military Necessity as Stewardship

Returning briefly to a point made in the introduction, the Civil War era Lieber Code, Article 14, defined battlefield military necessity as "measures which are indispensable for securing the ends of the war, and which are lawful according to the modern law and usages of war."[1] More than a century later, William V. O'Brien's classic definition is also useful,

> Legitimate military necessity consists in all measures immediately indispensable and proportionate to a legitimate military end, provided that they are not prohibited by the laws of war or the natural law, when taken on the decision of a responsible commander, subject to review.[2]

Although this book has rightly looked at various elements of *strategic* necessity, there is nevertheless a conceptual problem when military necessity as a tactical *jus in bello* concept and strategic necessity are confused. For instance, Michael Walzer's *Just and Unjust Wars* is often cited as discussing "necessity." But as has been made clear throughout this volume, Walzer's focus is on the necessity for political authorities to act in times of supreme emergency; that is, when faced with an existential threat. Against the backdrop of Israel's wars for survival in 1948, 1956, and 1967, Walzer writes, "The world of necessity is generated by a conflict between collective survival and human rights."[3] Walzer is just one of many authors who focus on the decisions that political authorities make when facing strategic-level disaster (*jus ad bellum*), in contrast to the issues of day-to-day battlefield fighting (*jus in bello*).[4]

As we have cast it, the best way to frame military necessity is in terms of the stewardship that military commanders must exercise under conditions of battle. In other words, military necessity is concerned with the practical aspects of battlefield scenarios that must be considered in light of the imperatives—practical and moral—of, broadly, the management of materiel, force protection, and victory.[5] Military necessity is not raw consequentialism. Rather, it works in tandem with the principles of proportionality and discrimination. Thus, it is best not to see the *jus in bello* criteria as a mere checklist, but rather see all three principles in overlapping, reinforcing, and informing relationship with one another. Military necessity, in our understanding, is not an occasion to abrogate the rules, nor to override morality in morally defensible ways. As Marc LiVecche's chapter makes clear: we assert that military necessity sometimes signals a moral imperative, not a moral paradox.

Military necessity as stewardship involves the careful management of those goods which have been entrusted, chief among them human life and property. Military necessity does include cost-benefit analysis, but, once again, this is not craven consequentialism. From both a Christian theological and a moral philosophical perspective, stewardship is essential. There are many biblical passages about kings and princes counting the cost before acting; and such is true in the battlefield principle of military necessity. The idea of stewarding manpower – the lives of sons and daughters of taxpaying citizens – is important for a republic. So too is the idea of good management of the limited financial and other resources that provide the armaments and tools necessary for battle. Military necessity as stewardship should matter greatly to a democratic society.

In practice, there are least four elements of military necessity that are important for stewardship. The first is *force protection*. Military commanders should be deeply concerned with protecting the lives and welfare of their own troops and this should influence the way they make decisions about battlefield activity. Moreover, military personnel themselves are citizens who, when at home, are husbands and wives, sisters and brothers, children, and parents. Soldiers are citizens, whether conscripted or voluntary, and we should safeguard their lives in accordance with the other *jus in bello* principles of proportionality and discrimination.

The second element is the principle of *economy of force*. Traditional military strategy has a set of principles such as not dividing one's forces, focusing one's forces in a certain area, and only using as much force as needed in a given domain. These are prudential considerations that have to do with husbanding one's resources. Such resources are not finite and are funded by the tax dollars of average citizens. These resources include more than military materiel, most importantly the lives of men and women in uniform. Moreover, when force is used in a deliberate but economical way, this will usually translate into far less destruction of the enemy, less damage to civilian infrastructure, far less collateral damage, and far less rebuilding.

Third, military necessity is also about *effectiveness*. Effectiveness means how well we get the job done. Historians are rightly concerned about how ineffective certain battlefield plans were in the First World War, such as at the Somme. In contrast, we laud the effectiveness of the battleplan in the 1991 Gulf War. Effectiveness is an important part of military necessity: Will this plan work? Will our actions be effective? In part, effectiveness and economy of force work in tandem, each hedging against the potential excess of the other. Effectiveness prevents economy of force from rendering military action indecisive. This is important. Decisive military action, purposefully aimed at the effective achievement of proper ends, helps to ensure that the fight that is right to fight is fought to win. Effectiveness should have a bearing on shortening wars and, having taken the fight out of them, preventing surrendered adversaries from going kinetic again. Ultimately, this, too, translates into far less waste and destruction. Effectiveness is a feature of moral accountability.

Fourth, military necessity is a link between battlefield activities and stated war aims (*jus ad bellum*) as well as a link between what happens now and the long-term accomplishment of war aims in *jus post bellum*. In other words, military necessity always has an eye on *victory:* victory in this engagement, victory in this campaign, and, ultimately, how this battlefield engagement advances long-term strategic victory. Too often arm-chair theorists want to use *jus in bello* to limit the activities and armaments of troops at the site of a specific tactical encounter. Military necessity reminds us that each battlefield is linked to larger campaigns in the grand strategy of the conflict.[6]

In short, military necessity is a critical stewardship principle that includes principles of *force protection, economy of force, military effectiveness*, and an *aim toward victory* that is important in providing a richer *jus in bello* that accompanies the principles of proportionality and discrimination. It is a moral criterion long recognized in the law of armed conflict, but that has fallen out of favor in recent years with just war scholars in the social sciences and humanities. The previous chapters in this book seek to redress this imbalance. The rest of this chapter points to opportunities for additional research and application.

Contemporary Issues for Additional Research

Military Necessity, Troop Protection, and Citizenship

A first area for additional research is the intersection of *military necessity* and citizenship, in particular the citizenship of troops while on duty. Too often recent philosophical discourse has wrapped itself up in seeking limits on harm to enemy combatants and non-combatants, to the detriment of our combatants. This is a moral issue faced by battlefield commanders and the duty for troop protection ascends up the chain of command, all the way to strategic guidance made by elected leaders and their immediate subordinates.

The Second World War provides exemplary cases as in the bloody island-hopping campaign in the Pacific, as well as later decisions made by President Harry S. Truman—and covered in this book. When one thinks about the will to fight and die of the Japanese, when many Japanese soldiers fought to the very last man, even when the battle was clearly lost, one must ask how much *American* life must also be sacrificed to end the extreme, evil, fight-to-the-death *Japanese* mentality? On battlefields such as Iwo Jima, Saipan, and Okinawa, commanders were wise and reasonable to consider every option, from aerial and naval bombardment to the use of area-clearing weapons such as flamethrowers. The alternative, hand-to-hand combat was practiced as well, but over time, it became clear that many Japanese were not fighting for a positive political settlement, but rather to fulfill a sort of feudal, honor-based demonstration of collective self-sacrifice. The degree to which it is possible to fight a just war justly when the adversary has no intention of doing so is a conundrum.

Marc LiVecche's chapter on President Truman and Hiroshima, as well as work by Eric Patterson on this decision,[7] focuses our attention on how the commander-in-chief has a responsibility to act within the framework of national security stewardship. Truman, who, unlike his predecessors, had fought in the trenches during the First World War I, approved the use of atomic weapons for reasons of military necessity. This is an action of stewardship of men and materiel, command, and preserving the lives of one's troops. And here is the key point. The Marines at Guadalcanal and Army forces who were ready to invade the homeland of Japan were not highly trained warriors and did not aspire to some mystical Samurai, or for that matter Teutonic, code when they woke on December 7, 1941. They were bakers, doctors, farmers, merchants, fathers, sons, husbands, and grandfathers; and their wives and children were pulled into the war effort whether working in factories or holding together the home front as America had to mobilize from a tiny army of just 300,000 troops to join the mighty Colossus that helped lead the allies to victory. *Military necessity* reminds us that President Truman and President Franklin D. Roosevelt before him, Prime Minister Winston Churchill, and others, had the responsibility to protect the well-being of their own civilian populations. Indeed, the first principle of national security is that leaders are responsible for the lives, livelihoods, and way of life of the citizenry that they represent. And U.S. *military necessity*, in terms of troop protection and all of its facets, comes into play on the battlefield. Consequently, we need more scholarship evaluating the way that the just war principles, and *military necessity* in particular, are informed by the fundamental assumptions of democratic government.

Military Necessity and Contemporary Rules of Engagement

A second area for expanded research is to consider *military necessity* and rules of engagement. When we think about the intersection of leaders' intent for going to war and the prudential categories of *jus ad bellum*, and how they are to be manifested on the battlefield, we rightly advance demands for proportionality and discrimination. But the question of whether such moral constraints are even feasible is often left begging.

For instance, proportionality requires that battlefield tools and tactics be commensurate to achieving battlefield objectives. It would be ridiculously disproportionate, and costly, to launch a nuclear missile to knock out a patrol boat. But commanders must consider how to reasonably protect the lives and property of legitimate non-combatants (e.g., civilians). In practice, this means protecting the many goods necessary for society to function, such as the electrical grid, water and sewage, houses of worship, schools, hospitals, and the like. From smart bombs to remote piloted aircraft with long loiter times, to satellite imagery, to basic military training that emphasizes fire discipline and other safety measures, much has been done to realize the full potential of these stewardship principles.

However, it does not make sense to say that the military response should be proportionate or discriminating if the troops or military in question do not have the tools, the training, the communications, and the direction that allow them to act in discriminating or proportionate ways. When we consider cases of armed humanitarian intervention over the past 30 years as a guide to future policy, we realize that most armed humanitarian intervention has not been performed by the highly trained, highly disciplined and well-equipped militaries of NATO and the Western alliance. Indeed, for much of the 1990s and early 2000s, the troops who ended up on the ground as UN peacekeepers came from among the poorest countries in the world. A recent ranking lists Nepal, Bangladesh, India, Pakistan, and smaller African countries such as Senegal and Ghana as providing the most troops for UN "blue helmet" missions.[8] In the first two decades following the Cold War, these troops did not have interoperability, and they still suffer from significant differences of training, weapons, and expertise. They did not, and do not, have the training or the tools at their disposal that Western militaries do. We need more thoughtful thinking on how *military necessity* requires preparation in terms of training, equipment, coordination, strategic focus, and troop protection.

At the same time, in those interventions that were carried out by Western powers, there have been serious questions about onerous rules of engagement that have restricted the effective use of force as well as putting more troops in danger for not being able to act with decisive force. For instance, European peacekeepers in the Balkans complained that they were politically constrained from stopping Serbian atrocities, but in other instances they were lightly armed and confronted by aggressive, well-armed forces. In Rwanda, in 1994, not only was UN peacekeeping leader Canadian General Romeo Dallaire told by his superiors to avoid engagement, but his mixed force of 5,000 blue helmets exhibited a dramatic range of quality, experience, and weaponry. So even if he had been given the go-ahead to intervene, he likely would not have had the fighting capacity to do so effectively.

Peacekeepers who are outgunned, poorly trained, and/or constrained by hyper-restrictive rules of engagement demonstrate a problem that occurs when *military necessity* is subordinated to risk-adverse conceptions of *proportionality* and *discrimination.* In truth, this breaks the bond between senior leaders and junior military personnel—that sacred bond that has to do with political leaders and commanders caring for the health and safety of their troops. These facets of *military necessity* have been on display when peacekeeping forces or humanitarians have landed in places such as East Timor and Afghanistan. In the case of East Timor in 1999, the Australian military, taking heed of rules of engagement that put peacekeeping troops at risk elsewhere, negotiated a solution: they would only lead the UN intervention if they were allowed a robust mandate that included elements of martial law and active measures of defense.

With 30 years of post–Cold War interventions now available as case studies, moral and policy reflection on military necessity and the rules of engagement is necessary. Indeed, the memoirs and reflections of America's generals who fought the Global War on Terror, such as Petraeus, Mattis, Franks, Dunford, Miller, and McChrystal, all identify debates over military necessity and the rules of engagement. Looking forward, particularly as military and political leaders consider involvement in gray zone operations (e.g., military operations other than hot war), more thinking on military necessity and the rules of engagement will continue to be needed.

Military Necessity as National Security Stewardship

The third area for more scholarship is in an area that we call national security stewardship. One of the first times that I thought of senior leaders as the stewards of the people in their charge was when I was invited to speak at U.S. Central Command alongside James Turner Johnson. Over lunch with General James Mattis, we discussed the ethics of dealing with imprisoned insurgents. At the time, the Obama Administration had made many public pronouncements about its desire to get out of Iraq and Afghanistan as quickly as possible. One of the outstanding issues was what to do with the criminals and terrorists behind bars in prisons across Iraq. The Obama administration's catch-and-release policy, not just in Iraq but also on the high seas in dealing with Somali pirates and elsewhere, resulted in fighters returning to the fight, after they had experienced a sometimes quite short period of confinement. We had seen what happens when a prison was turned over to the Government of Iraq: in some cases, the cells were unlocked and everyone walked out, including murderous Iranian agents who had killed American troops.

General Mattis worried, privately and later publicly, over the ethics of releasing a terrorist from custody, who could return just a few days later to murder a young American Marine. How could it be morally right and responsible policy for the United States to release a guilty and unrepentant murderer whose purpose is to renew the fight? Mattis ultimately stepped down from his position, ostensibly due to this and other disagreements with the Obama Administration. This situation, which had analogies in Afghanistan as well, reminded me about the biblical principle of stewardship and counting the cost, whether building a tower or making an investment, and how different General Mattis's concern for the average soldier and Marine was from leaders throughout history who had very little concern for the lives of the junior-most people under their command. The active ethic under consideration at the operational level is, of course, *military necessity*: balancing mission, force protection, the tools that are at hand and an accounting of the resources available, and consideration of a variety of operational possibilities given real-world constraints. The term that I have introduced to get a handle on this is *national security stewardship*.

One more point is in order here. Much of the world seems to think *military necessity* as national security stewardship does not apply to the United States. Surely that there are no strategic limits to our ambitions—no constraining national debt, no limited natural resources, no dearth in the availability of manpower, armaments, and the like. Isn't the so-called American way of war to develop a massive military industrial complex and slam it down on our adversaries? How can one be concerned about cost savings for such an enterprise? Moreover, when that behemoth is unleashed, shouldn't there be a lack of restraint in all areas, including the sacrifice of our troops? Are good commanders not willing to spend the lives of their troops? Of course, this is all nonsense. Thinking about limits, restraint, and the human and material costs of war is a practical necessity—even for the United States. Even while the U.S. gross domestic product (GDP) is more than that of our chief adversaries—Russia and China—combined, and even as our defense spending outstrips their combined expenditures, the well is not bottomless. And while it is true that good commanders will be willing to spend the lives of their troops, it is also true that they must never waste them. As this last point makes plain, then, stewardship is more than a practical necessity—it, too, is a moral imperative.

Military Necessity and Operational Art

Finally, *military necessity* is the realm of operational art at the strategic level as political leaders and their advisors make broad decisions that lead to national security policies and then military operations. When those operations are planned and commence they are within the realm of *military necessity;* and thus, if our just war thinkers are abandoning or avoiding this important principle, then what they are demonstrating is an utter lack of regard for the ethics of how we train and prepare soldiers and how soldiers make decisions on the battlefield. Of course, this may be due to the fact that so few of the louder voices among philosophers, theologians and other "theorists" have any sense about what the military is actually like in the first place. Few of them have volunteered for military service, but rather speak from the comfort of academic departments. Perhaps the most egregious example of this is the contrast between President Truman and the young Oxford philosopher, Elizabeth Anscombe, who later damned him for his decision to use the atomic bomb. As LiVecche recalls in his own discussion of Truman's decision, Anscombe, in 1956, famously—or infamously—challenged Oxford University for presenting President Truman an honorary doctorate for not only his magnificent leadership at the end of the Second World War but also for his leadership in the establishment of NATO, the United Nations, the implementation of the Marshall Plan, and other robust efforts at international peace and security. Anscombe concluded that Truman's decision to use the atomic bomb was immoral. Her evaluation of an award to Truman can be summed up in her statement, "If you do give this honor, what Nero, what Genghis Khan, what Hitler, or what Stalin will not be honored in the future?"[9]

Of course, the difference between the informed practical philosophy of each of these two individuals is telling. Anscombe, a student of Wittgenstein, largely spent her life in the academy, including during the Second World War. Truman was the only U.S. president to have fought in the First World War. Truman knew warfare. He knew how hard it could be and he knew the value of the men in the trenches fighting the war. He had been one. His friends were doughboys and he famously stayed in touch with Company D of the 128th Field Artillery (35th Division). He had gone through years of modest training with the National Guard, then accelerated artillery preparation and officer development prior to deployment. His philosophy of wartime leadership, as president, was therefore informed by his training and experience, as well as his legendarily massive reading on history and warfare.

What President Truman had not learned by experience and training he gleaned from the outstanding cadre of advisors that surrounded him. He would have known the costs in lives and infrastructure that the Japanese would have suffered had any of the other militarily viable alternatives to the bomb were used. He would have learned about the extraordinary loss of life being suffered every day by innocent captives living in lands under Japanese occupation. This was utterly unlike Anscombe's academic situation. The difference should make us think not simply about lived experience but also about operational art. Operational art is defined as: "The employment of military forces to attain strategic and/or operational objectives through the design, organization, integration, and conduct of campaigns, major operations, and battles."[10] Truman was thinking at the strategic level and was "staffed" by some of the greatest operational "artists" of the day, most notably General George C. Marshall as Army Chief of Staff and the outstanding Secretary of War, Henry Stimson.

To date, little work has been done considering the intersection of the ethical concept of military necessity and operational art. This is likely because the linkage is starkly obvious. However, the fact that just war thinkers have largely neglected military necessity since the Second World War suggests that a recovery of the principle and an analysis of its effect on operational art is a fruitful avenue for additional research.

In conclusion, the return of military necessity is desperately needed and yet is a misnomer. On battlefields everywhere, military necessity has never left. But its neglect in recent decades by scholars, including the editors of this book, calls us to better consider the vitality and usefulness of the concept as we consider just war statecraft and national security stewardship in the years ahead.

Notes

1 Richard Shelly Hartigan, *Lieber's Code and the Law of War* (Chicago: Precedent Press, 1983).
2 William V. O'Brien, *The Conduct of Just and Limited War* (New York: Praeger, 1981), 9.
3 Michael Walzer, *Just and Unjust Wars*, third ed. (New York: Basic Books, 2010), 251–252.
4 In a helpful book, Michael Gross uses the principles of military necessity, effectiveness, and proportionality to talk about a two-step distillation process for making decisions about battlefield necessity, first by deciding what is necessary and then second, taking a clear-eyed look through the lens of humanitarian principles at such actions. Gross goes on to suggest an evaluation or feedback for new weapons and tactics, considering military necessity use, limits of humanitarianism, and then re-evaluating the effectiveness of necessity and the limitations: Did it work as expected? Do we need new regulations or changes as we go forward? Clearly this is intelligent and ethical, but this reflective approach is the purview of military and civilian leaders, not lieutenants and soldiers under fire. See Michael Gross, *Soft War: The Ethics of Unarmed Conflict* (Cambridge: Cambridge University Press, 2017).
5 Eric Patterson, *Just War Thinking* (Lanham, MD: Lexington Books, 2007).
6 James Dubik makes a not dissimilar argument in the introduction to his *Just War Reconsidered* (University of Kentucky Press, 2016). He argues that we need to focus more attention on the "gap" between the *jus in bello* direction that senior leaders must provide and the *jus in bello* activities going on at the tactical level; the former needs more attention.
7 Eric Patterson, *Just American Wars: Ethical Dilemmas in U.S. Military History* (New York: Routledge, 2019).
8 "Largest Contributors of Troops to UN Peacekeeping." Statista. Available at: https://www.statista.com/statistics/871432/largest-contributors-of-troops-to-united-nations-peacekeeping/. Accessed May 1, 2023.
9 Quoted in John Schwenkler, *Anscombe's Intention: A Guide* (Oxford: Oxford Academic Books, 2019), 23.
10 U.S. Air Force. "Operational Art," in *Air Force Vocabulary*. Retrieved at: https://www.nationalmuseum.af.mil/Visit/Museum-Exhibits/Fact-Sheets/Display/Article/198085/air-force-vocabulary-j-p/.

Index

Note: Page numbers followed by "n" denote endnotes.

For Product Safety Concerns and Information please contact our EU
representative GPSR@taylorandfrancis.com
Taylor & Francis Verlag GmbH, Kaufingerstraße 24, 80331 München, Germany

www.ingramcontent.com/pod-product-compliance
Lightning Source LLC
Chambersburg PA
CBHW060307220326
41598CB00027B/4263

* 9 7 8 1 0 3 2 4 8 7 1 2 0 *